Animals of the Antarctic

the ecology of the Far South

MANNERING

Animals of the Antarctic

the ecology of the Far South

by Bernard Stonehouse

BOOK CLUB ASSOCIATES
LONDON

Wave-worn iceberg. This small fragment of berg began on land as glacier or shelf ice, and has melted slowly during several summers at sea.

Previous page: Emperor penguins, some with eggs on their feet, huddle in early spring sunshine. Dion Islands, Marguerite Bay, southern Antarctic Peninsula.

MICKLEBURGH

The publisher gratefully acknowledges the help and advice given by the Scott Polar Research Institute, Cambridge and the British Antarctic Survey.

Maps by Cartographic Services (Cirencester) Ltd

Index compiled by Dorothy Frame

This edition published 1974 by
BOOK CLUB ASSOCIATES
By arrangement with Eurobook Ltd.

Printed in the Netherlands

Introduction

Far to the south of cities and civilization lies Antarctica, a white continent broader and wider than Australia or the U.S.A. Loftiest of the world's seven continents, with a heartland colder than anywhere else on earth, Antarctica is centred about the south geographic pole and spreads north in every direction toward the Antarctic Circle. Frosted crazily like an eccentric wedding cake, it carries nine-tenths of the world's permanent ice, in an irregular layer of immense thickness. The high central dome, with ice up to four kilometres thick, hides mountain chains longer than the European Alps, and ice-choked valleys plunging far below sea level. Less than a twentieth of its ancient crust is visible, as peaks, mountain ranges and islands thrusting through the ice cover.

Antarctica for millions of years was a temperate continent, with forest and grassland, lakes, swamps, rivers and alpine meadows. Like every other continent it had a fauna of walking, flying, crawling, climbing and burrowing animals, including reptiles, mammals and birds. Only within the last four or five million years – the period of man's emergence – has it chilled to a wilderness of ice, with mineral-starved soils, a cold desert flora of simple plants, and a fauna dominated by tiny mites and insects. A book on the modern animals of continental Antarctica alone would be brief and rather drab; in all the continent at present there is no land animal bigger than a housefly, or more colourful than dull grey felt.

But by a curious paradox Antarctica is ringed by the world's richest and liveliest ocean, an ocean dotted with colourful islands and swarming with energetic, splendidly adapted marine life. Continent, ocean and islands together make up the Antarctic region, and the animals of the Antarctic introduced in this book are mostly marine animals. There are delicate floating animals of the surface waters, fishes, elusive squid which every animal but man seems able to catch, seals, whales, and a host of penguins, petrels, skuas and other sea birds. These are the true inhabitants of the Antarctic region, animals completely at home in an environment where man must always be alien.

Today it is impossible to think of animals without concern for man's impact upon them. Even at the far southern end of the world, where there is seldom more than one man to every ten thousand square kilometres of wilderness, animals face threat of pollution, exploitation, and other lethal influences. Pollution of several kinds has already spread to the Antarctic from northern civilization. Adelie penguins, crabeater seals and Nototheniid fishes have been found to carry organo-chloride pesticide residues in their tissues, and the snow of the Antarctic ice cap, which might once have stood as a symbol of purity, contains fission products from thermonuclear tests of the 1950s and 1960s. Exploitation devastated stocks of Antarctic fur seals and elephant seals in the nineteenth century, and in the present century has reduced numbers of Antarctic whales to dangerously low levels. Now the future looks grim for crabeater seals—the silver-white seals of the pack ice—for man is beginning to find that he needs their skins, and crabeater seals on the pack ice of the high seas have no protection in law.

But a more general threat hangs over practically every bird and mammal in this book. Man may not at present want their fur or feathers or the ground they breed on. He wants nothing less than their food. Already he has begun to harvest the plankton—the crustaceans and larval fish which, seasonably abundant in surface waters, directly or indirectly feed practically every Antarctic animal. If he harvests it wisely, he might take forty million tons of palatable plankton from the sea annually without over-fishing, and still leave plenty for the birds and mammals which unwittingly compete with him. But the plankton, like the crabeater seal, has no national or international law to protect it. Anyone may take it, and the greatest immediate reward goes to those who take as much as possible, as quickly as possible. Within a very few years, modern technology could make possible annual catches of two, three or four times the sensible, sustainable level, and man would be destroying the final and greatest bounty which the southern polar region has to offer him.

Though international expeditions and tourist parties now visit the far south each year, still very few men and – monstrously – fewer women have had a chance to experience the glowing beauty of the Antarctic region, and enjoy the company of its animals. This is unfortunate, for it means that very few people have a stake in the Antarctic, or feel committed to defending the forces which threaten it. I hope that this book will help to spread an enjoyment of the Antarctic, and show what the far south has to offer mankind.

Bernard Stonehouse
Forgandenny
Perthshire

The Antarctic Region

MANNERING

SWITHINBANK

Cape Agassiz, an ice-covered headland on the east coast of Antarctic Peninsula, surrounded by the Larsen shelf ice.

Previous page: Ridges and valleys of an "oasis" area, southern Victoria Land. These valleys, once ice-filled, are now a dry desert. Pale, horizontally-bedded Beacon Sandstones contrast with dark, columnar intruded dolerites.

The Antarctic is the region of cold surrounding the South Pole. Geographically it includes a continent, a broad expanse of ocean, and a scattering of small archipelagos and islands, with a total area of about 45 million square kilometres (or one-eleventh of the earth's surface). The continent is Antarctica itself, fifth largest of the world's seven continents and last to be discovered. The ocean might sensibly be called the Antarctic Ocean, though few cartographers seem anxious to use the term: it is the farthest zone of the great southern ocean which rings Antarctica like a moat. The islands include the snow-capped archipelagos of the Scotia Arc (South Shetland, South Orkney and South Sandwich Islands), isolated South Georgia, Bouvetøya, Heard Island and Îles Kerguelen, and a host of smaller islands standing under the lee of continental Antarctica (page 24).

The name "Antarctic" was coined long before man had ventured far south. At first it implied no more than the opposite of "Arctic"—the north polar region dominated by Arktos, the Great Bear constellation—and referred in a vague way to the whole unknown southern end of the world. "Antarctic" became a synonym of remoteness, opposition, contradiction, a term to apply to a political opponent, or a philosophy remote from one's own. World maps of the sixteenth and seventeenth centuries showed a continent surrounding the South geographic Pole . This was not Antarctica; it was *terra australis incognita,* a curious imaginary landmass which no one had seen, but which early geographers believed linked the southern tips of Africa and the Americas and filled the polar region. *Terra australis incognita* gave its name to Australia, the first real continent to be carved out of it by Dutch and British voyagers of the seventeenth century. Antarctica, the far southern continent, remained unknown until the mid-nineteenth century. First seen by man in 1820 and first recognized as a continent some thirty years later, Antarctica took its name almost by default from the far southern region which it dominates.

SWITHINBANK

The sunlit peaks of Eternity Range, seen over the plateau of southern Antarctic Peninsula.

Fur seals basking in weak summer sunshine on the shores of South Georgia.

TICKELL

The Antarctic continent

As the astronaut sees it, Antarctica is a white, comma-shaped continent surrounding the South Pole, with an area of roughly 14 million square kilometres. Most of this is ice, a vast permanent sheet of it averaging 2,000 metres thick, which spreads like a ragged cloak and hides the true contours of the underlying continent. Antarctica beneath the ice is a much smaller landmass, possibly of some 7 million square kilometres, with a neighbouring archipelago of mountainous islands separated by deep channels and land-locked seas (page 17). Only about one-fortieth of its land is actually visible as mountains or coasts.

The head of the white comma is Greater Antarctica, a semi-circular or elliptical mass. Much of its rim lies close to the Antarctic Circle ($67\frac{1}{2}$°S.), bordering the southern Atlantic, Indian, and western Pacific Oceans. The tail of the comma, 1,200 kilometres long, emerges from the flank of Greater Antarctica bordering the south-eastern Pacific Ocean, and points like a finger in the direction of South America. The root of the tail is Lesser Antarctica; the tail itself, formerly called Graham Land, O'Higgins Land or the Palmer

YOUNG

Low precipitation in southern Victoria Land has produced "dry valleys"—desert areas without ice—on the seaward flanks of the coastal mountains.

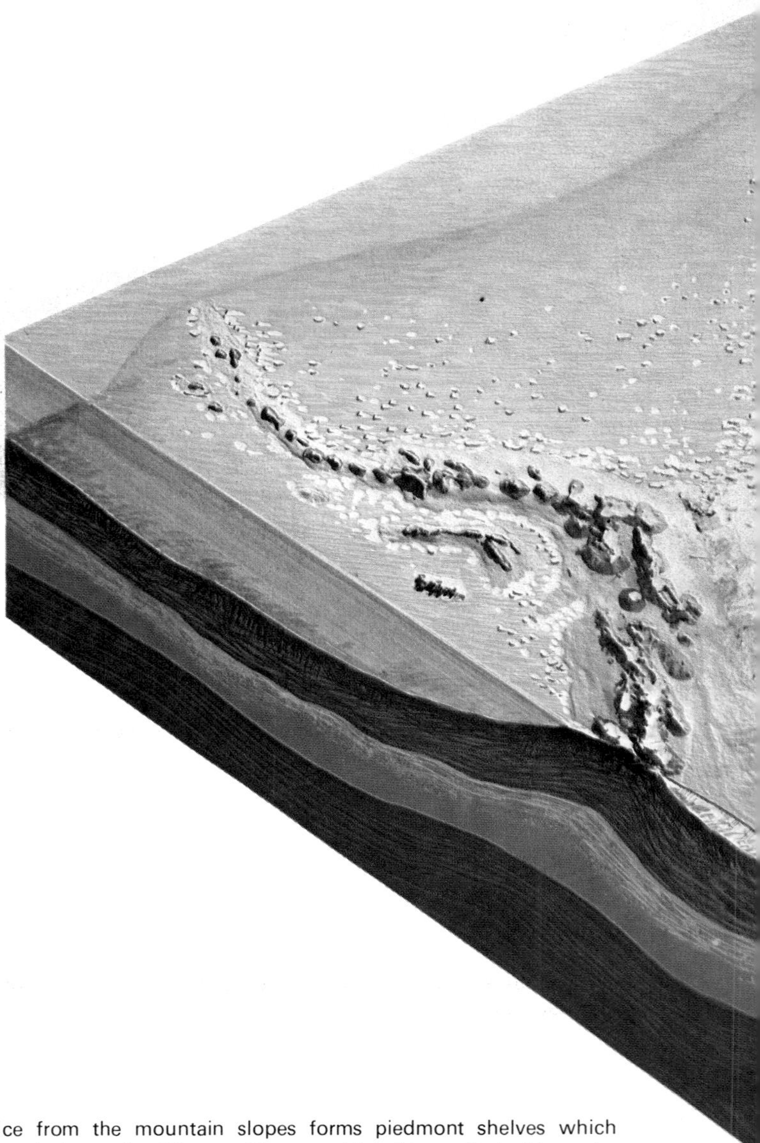

SWITHINBANK

Ice from the mountain slopes forms piedmont shelves which reach the sea in cliffs over 30 metres high. Mount Paris, Alexander Island.

Peninsula, is now called Antarctic Peninsula. The comma surface is white and undulating, broken here and there by jagged mountains which thrust through holes in the ice mantle. Antarctic Peninsula is a crescent of alpine mountains, topped by an ice plateau and carved by massive active glaciers. The coastal rim of the continent is mostly ice cliff. At only a few points has the mantle edge frayed and rolled back, exposing a raw shoreline of rocky headlands and beaches.

Antarctica with its ice cap is by far the highest continent, averaging over 2,000 metres in elevation. Its tallest mountain, Vinson Massif in the Sentinel Range of Lesser Antarctica, rises to 5,140 metres, and there are many peaks over 3,000 metres high. Greater Antarctica's main ice dome rises to 4,200 metres or more; Lesser Antarctica is lower, with a broad, undulating plateau averaging 2,000 metres high. Sledge traverses and survey flights of the last two decades have mapped much of the ice surface,

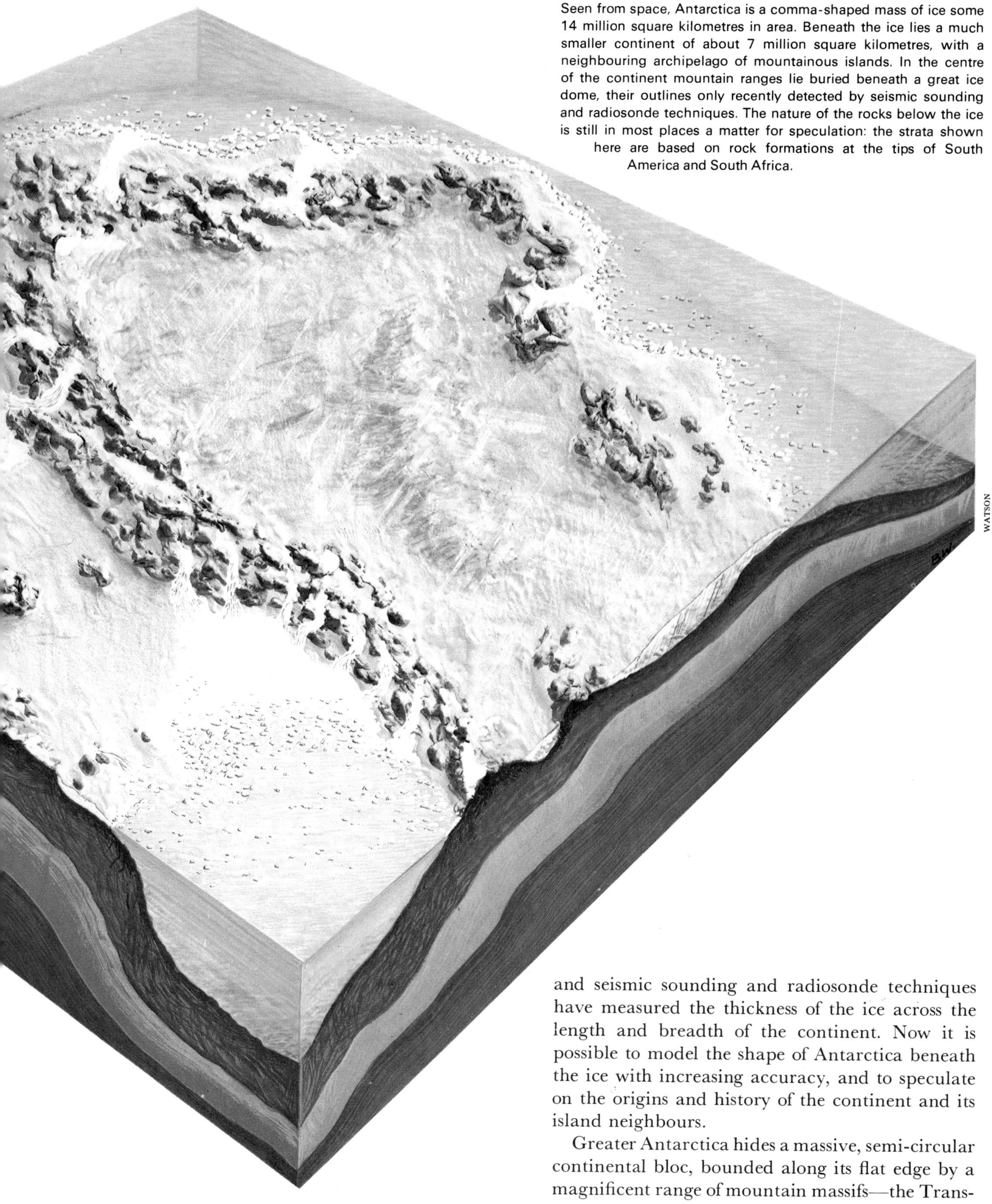

WATSON

Seen from space, Antarctica is a comma-shaped mass of ice some 14 million square kilometres in area. Beneath the ice lies a much smaller continent of about 7 million square kilometres, with a neighbouring archipelago of mountainous islands. In the centre of the continent mountain ranges lie buried beneath a great ice dome, their outlines only recently detected by seismic sounding and radiosonde techniques. The nature of the rocks below the ice is still in most places a matter for speculation: the strata shown here are based on rock formations at the tips of South America and South Africa.

and seismic sounding and radiosonde techniques have measured the thickness of the ice across the length and breadth of the continent. Now it is possible to model the shape of Antarctica beneath the ice with increasing accuracy, and to speculate on the origins and history of the continent and its island neighbours.

Greater Antarctica hides a massive, semi-circular continental bloc, bounded along its flat edge by a magnificent range of mountain massifs—the Trans-

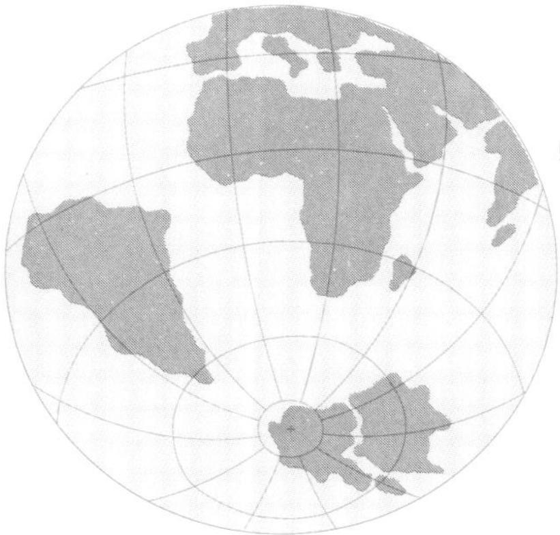

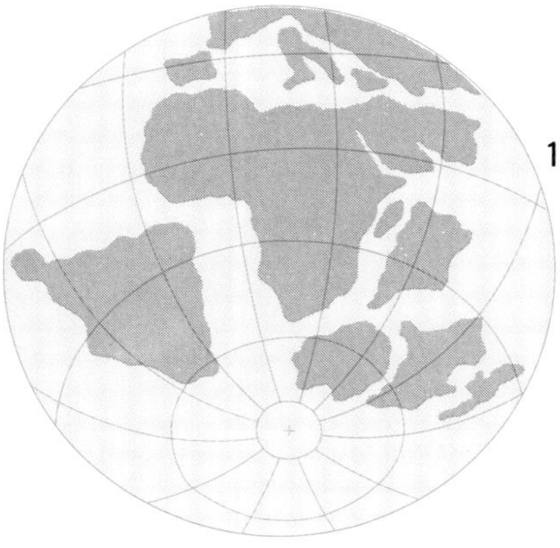

YEARS IN MILLIONS	*WORLD EVENTS*	*ANTARCTICA*
0	**PLEISTOCENE AND RECENT** Start of Arctic Ice Age. Spread of Man.	Glaciation complete.
	PLIOCENE Man-like apes in Africa. Cool temperate climate in Britain. Andes and Rocky Mountains upthrust.	S. Sandwich Islands formed. Ice cap spreading to coast. Floating ice cooling Polar sea.
	MIOCENE Continents assuming present positions. Elevation of Alps, Himalayas, Andes. Forests change to grasslands. First appearance of man-like apes in Asia.	Alpine mountains elevated. Ice well established in highlands. Penguins, primitive whales (Seymour Island).
	OLIGOCENE Start of Alpine mountain building. British and North American flora sub-tropical. First appearance of seals, sea-lions, albatrosses.	Possible formation of permanent ice in highlands. Alpine mountains starting to appear.
50	**EOCENE** World climate continues cooling. Australasia and Antarctica separate. Elevation of continents. First appearance of whale-like mammals, penguins, gulls.	Isolation begins. Southern beech forests (S. Shetlands).
	PALAEOCENE Southern Atlantic and Indian Oceans widen. World cooling begins. Flowering plants and mammals begin to dominate land.	Mountains of Lesser Antarctica, S. Georgia and S. Shetlands elevated. Southern beech in coastal areas of continents.
	CRETACEOUS Extinction of many reptiles.	
	World-wide temperate climates, chalk seas, low-lying continents.	Mountains of Dronning Maud Land lifted by intrusion.
	Development and spread of flowering plants.	Cool temperate sea in Antarctic Peninsula area. Molluscs, crustaceans, worms, fishes.
100	Small mammals diversify.	
	Americas begin to swing westward.	
	Antarctica and Australasia move south and east in company.	
	First flying birds appear.	Ammonites, fish remains (South Georgia).
	Flowering plants appear.	Lake deposits (South Orkneys).
	JURASSIC Rifts appear in Laurasia (north) and Gondwanaland.	Volcanic activity, possibly local high-level glaciation.
	Widespread basalt lava flows.	
150	Ferns, conifers predominate.	Belemnites in warm sea (Alexander Island), Ellsworth Land.
	Dinosaurs and other reptiles dominant.	Volcanic activity, Antarctic Peninsula.
	First mammals appear.	Cycads, conifers, bivalve molluscs, snails, beetles. Fishes in warm-temperate lakes.

antarctic Mountains. For nearly 2,000 kilometres this range is exposed along the flank of Victoria Land, where it supports and holds back the edge of the Greater Antarctica ice dome. Glaciers pour between its mountains, forming broad piedmont ice shelves at the northern end, and merging into the vast, floating Ross Ice Shelf closer to the pole. Less than 450 kilometres from the pole, the mountains are overridden by the thick ice of the polar plateau. They skirt the pole under the mantle, and reappear as the Pensacola Mountains of Coats Land on the opposite side of the continent.

Several other mountain ranges lie more or less hidden under the ice mantle of Greater Antarctica. The smooth surface of the ice dome completely obliterates a vast range more than 3,000 metres high and 1,300 kilometres long—a range more extensive than the European Alps. Enormous coastal ranges border the Atlantic and Indian Ocean sectors of the continent, appearing only as isolated peaks and rock escarpments partly hidden under tumbling ice-falls. Between the ranges are low-lying plains and extensive basins, overlain by some of the thickest ice on the continent. Amundsen-Scott, the permanent U.S. base at the South Pole, stands 2,800 metres above sea level, and about the same height above bed-rock (page 24). But for the ice, the geographic South Pole would lie in a shallow basin close to the foothills of the Transantarctic Mountains. Deeper basins underlie other parts of Greater Antarctica, delving well below sea level and supporting ice over 4,000 metres thick.

Geologically, Greater Antarctic is a rock platform composed of a stable shield of ancient metamorphic and intruded rocks, overlain by later sediments. The shield or platform, dating from Precambrian and early Palaeozoic times, is over 500 million years old. It appears as exposed rock faces in Dronning Maud Land, and in the coastal mountains of the Indian Ocean sector; it underlies the length of the Transantarctic Mountains, and probably forms the foundation of the whole of Greater Antarctica. Resting on the platform is a sequence of younger sedimentary rocks of late Palaeozoic to mid-Mesozoic age (200 to 400 million years old). Forming the so-called "Beacon Series" of sediments, these include marine muds, shales, estuarine and fresh-water deposits, thick coal measures, and desert sandstones. Between the flat-bedded sediments are broad bands and lenses of dark, fine-grained dolerite, a rock injected by volcanic action long after the sediments were laid down. More resistant to erosion, the dolerite caps

SWITHINBANK

Stratified ice melts in summer sunshine, exposing bare rock to the shattering forces of polar weathering. Near Novolazarevskaya, Dronning Maud Land.

the softer sedimentary rocks and forms prominent horizontal platforms. The Beacon Series, roughly 2,500 metres thick, is most clearly exposed in the vast uplifted blocs of the Transantarctic Mountains. These were raised by earth movements probably during the mid-to-late Tertiary, i.e. during the last 35 million years.

The ancient Precambrian platform and the Beacon sediments together tell of the remarkable travels and history of this sector of the Antarctic continent. Firstly, they show unmistakably that Greater Antarctica is largely a fragment of Gondwanaland, a great southern supercontinent which, originally including southern Africa, India, eastern Australia and South America, split during the late Jurassic or early Cretaceous periods some 200 million years ago (page 18). Each of the fragments, including Greater Antarctica, is made up of similar rocks in similar sequence, and shows a common history of deposition, erosion, glaciation, and other geological events. The fragments of Gondwanaland drifted slowly to their present positions after the break-up, Greater Antarctica remaining linked with Australia probably until the early or middle Tertiary, up to 50 million years ago.

Secondly, the rocks of Greater Antarctica show that, during its association with Gondwanaland and later wanderings, the continent experienced a wide range of climatic conditions, from an earlier ice age during the Permian and Carboniferous (250 to 300 million years ago) to hot, dry desert conditions in the Mesozoic. The Beacon sediments

Snow algae, minute unicellular plants, colonize the surface of melting snow on the Antarctic coast, forming vivid patches of red, yellow or green.

Far right: Nimrod Glacier, a slowly moving cascade of ice, which descends from the plateau of greater Antarctica to merge with the Ross ice shelf. Transantarctic Mountains, southern Victoria Land.

Midnight sunshine casts long shadows—16 to 20 kilometres long—over the jagged peaks of southern Antarctic Peninsula. Fallières Coast, Graham Land.

LEWIS SMITH

SWITHINBANK

have yielded a fossil harvest of ferns, cycads and horse-tails, fossil shrubs and trees, fresh-water crustaceans, and—some exciting recent discoveries—the bones of amphibians and reptiles. Later fossil deposits, probably but not certainly of early Tertiary age (i.e. up to 70 million years old), suggest that beech trees and palms were once prominent on Antarctic coastal lowlands. Forests of southern beeches once grew in the hinterland of McMurdo Sound, now a barren desert where even mosses and lichens survive only with difficulty.

Lesser Antarctica's twisted, contorted alpine mountains contrast sharply with the monolithic blocs of Greater Antarctica, and their history is totally different. Exposed along the flanks of Antarctic Peninsula, they extend from the peninsula tip toward the heartland of Lesser Antarctica, where they disappear under the ice of Ellsworth and Byrd Lands. Under the ice the mountain chain spreads and breaks into blocs, separated from each other by steep-sided, ice-filled canyons which descend over 1,000 metres below sea level. Peaks and shoulders of the mountains break through the ice mantle, forming the scattered, isolated outcrops and ranges of the sub-continent. Off the north-western flank of the Peninsula a second line of peaks forms the Palmer Archipelago, Biscoe Islands and Adelaide Island, continuing north and eastward in the long chain of the South Shetland Islands. Submarine contours link the South Shetlands with the outlying South Orkney Islands, then sweep northward and west to include the South Sandwich Islands and South Georgia, and ultimately to join the submarine shelf ringing the tip of South America. This chain of islands and submarine ridges is the Scotia Arc, a region of stress in the earth's crust with a long history of volcanic action, earth movement, and crustal folding.

The mountains of Lesser Antarctica are metamorphosed sedimentary rocks, interbedded with volcanic ashes and lavas, and underlain with massive intrusions of dolerite and granite. The rocks include thick bands of greywacke, which was deposited as a fine grey mud in deep water and subsequently squeezed, consolidated and baked rock-hard by intense heat. The mud was laid down in long troughs or geosynclines, temporary downfolds in the earth's surface which formed repeatedly along the same zone of crustal weakness. The Andes of South America and the alpine mountains of eastern Australia and New Zealand were formed in similar geosynclines, and bear a strong family

BONE

Icebergs are formed from glacier and shelf ice which move constantly seaward and break off in deep water. Gravel and rocks trapped in the ice are strewn over the sea bed as the bergs melt.

MICKLEBURGH

The tumbled snout of North East glacier calves constantly throughout the summer, shedding small icebergs into Marguerite Bay, Antarctic Peninsula.

resemblance to those of Lesser Antarctica and the Scotia Arc. Lesser Antarctica's mountains represent many cycles of deposition, folding, uplift and erosion from Precambrian times onward. When earth movements lifted the sediments far above sea level, the land was colonized by plants and animals. Ferns, conifers, extensive swamp vegetation, beetles, flies and molluscs are found in the Mesozoic fossil record of Lesser Antarctica, though hot volcanic ash and lava, coupled with folding and twisting due to earth movements, displaced and destroyed many of the organisms and their remains.

During the late Cretaceous or early Tertiary a series of violent upheavals, accompanied by massive injections of molten rock, lifted both the archipelago of Lesser Antarctica and the islands of the Scotia Arc. Formerly more extensive, they suffered many millions of years' weathering and erosion under temperate climates, and huge tracts of land slipped back into the ocean as massive faulting and folding continued. Rifts opened in the sea bed and on land, pouring out ash and molten rock. Volcanic and intrusive action continued well into the middle and late Tertiary (i.e. the last 35 to 40 million years), producing new islands in the South Shetlands group, the whole new chain of South Sandwich Islands, isolated Peter I Øy and Bouvetøya, the Edsel Ford Range bordering the Ross Sea, and probably many other mountains now under the Lesser Antarctic ice sheet.

Deception Island in the South Shetlands, Bouvetøya, and several of the South Sandwich Islands are still volcanically active, and the whole geosynclinal region of Lesser Antarctica remains a lively earthquake belt. Greater Antarctica is more sedate, with only two very restricted areas of volcanic activity—Mount Erebus on Ross Island, and Mount Melbourne in northern Victoria Land. Escaping gases and mineral solutions provide patches of locally warmed soils, which mosses and microscopic animals colonize to their advantage.

The Antarctic ice cap

Through most of the earth's history polar regions have been free of permanent ice, and the climatic gradient between polar, temperate and equatorial regions has usually been less marked than it is today. The poles have always received their sunshine at a low angle, and so remained cooler than the equator, but they have usually managed to remain free of permanent ice. Why are they now ice-bound? We do not know the full answer, but many lines of evidence indicate that the polar ice caps developed after a long period of world cooling. Between the middle of the Mesozoic era (150 million years ago) and the start of the Pleistocene (three million years ago), the mean surface temperature of the earth is believed to have fallen from about 20°C. to a little above 10°C. Polar regions cooled more than equatorial regions; permanent ice began to form, and eventually both southern and northern ice caps developed.

The long spell of world cooling may have been due to changes in the distribution of sea and land in the polar regions. Many climatologists now believe that polar regions become frigid only when they are surrounded by continental land masses or land-locked seas, but remain warm when oceans extend to the poles. With the poles in mid-ocean,

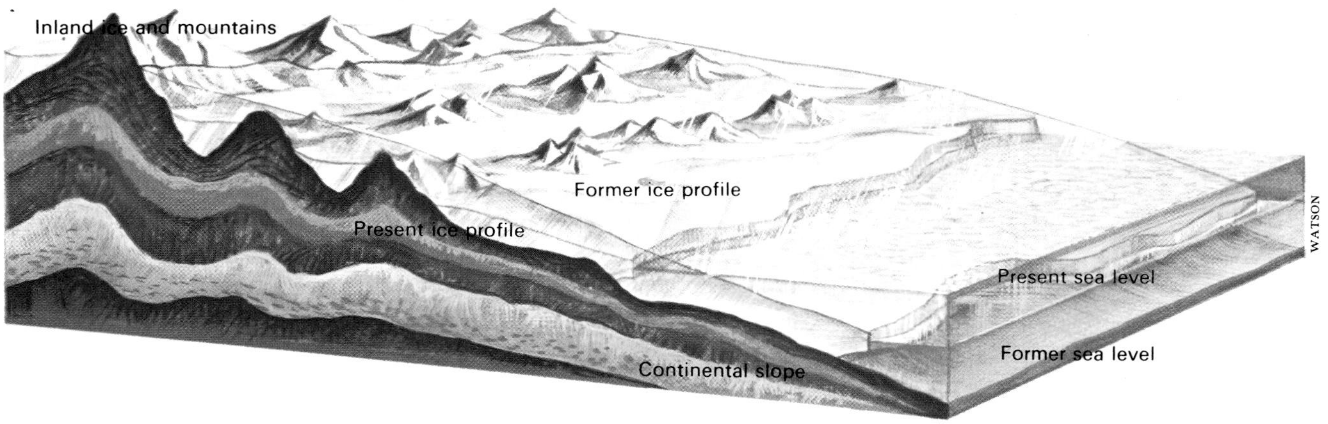

How changing sea level determines the size of the ice cap. A fall in sea level (caused by extensive glaciation elsewhere on earth), exposes a ring of new coastal land; the ice advances to cover it, and the old coast is buried deeper. When sea level rises, the new coastal ice breaks back and floats away, and the original profile of the coastal slopes is restored.

north-south currents carry warmth from the tropics into high latitudes, dispersing seasonal ice before it has time to become permanent. With the poles on land, or isolated from the oceans by land barriers, the warmth cannot penetrate and winter ice becomes permanent. If the land is high, the ice has an even better chance of persisting and developing a zone of intense cold about its perimeter. Up to about 70 million years ago the poles may well have been located on low ground or in the ocean. When the South Pole came to coincide with the Antarctic land mass, and when mountain building raised huge tracts of the polar continent high above sea level, the southern regions began to chill and the whole earth felt the effects of their cooling.

The southern ice cap probably began as scattered areas of snow high among the new mountain ranges. Their shiny white surface reflected back radiation which had previously warmed the earth, and the snowfields became permanent. New snowfields, forming in the newly-elevated Pyrenees, Himalayas and European alps, helped to cool the world further. The Antarctic snowfields united to become ice sheets, which spread from high land to low land and gathered for their advance over the continental plains. Each advance covered more ground with ice, caused more solar radiation to be reflected into space, and brought further cooling to the polar region. By the mid-Pliocene epoch, some five to six million years ago, cold temperate climates still prevailed close to sea level in both polar regions, but the cold was biting deep. In the far north coniferous forest and tundra were starting to replace broad-leaved forests along the Arctic shore, and ice was forming in high country. We do not know what was happening in the far south, for most of the evidence lies under the ice sheet or was scraped from the continent as the glaciers advanced. But we do know, from the evidence of mud samples of known age deposited at the bottom of the Southern Ocean, that the first Antarctic glaciers reached the coast and began to form floating tabular bergs some four million years ago. From that time, glacial mud and angular, ice-shattered rock fragments began to spread like an ever-widening carpet across the southern sea bed: the Antarctic ice age had begun.

In northern latitudes the ice age began slightly later. Two to three million years ago permanent ice formed in the cold heartland of Europe and Asia, and covered the mountains of Greenland, Iceland and North America. About one million years ago the first ice sheets spread to the valleys of North America and Europe (the Nebraskan and Donau glaciations). Then followed the four major glacial phases of Europe—the Günz, Mindel, Riss and Würm of Alpine glaciologists—which, together with their long, warmer interglacial spells, occupied most of the last million years. Each glacial phase saw the advance of ice sheets across the plains of Europe and North America, and probably across both eastern and western Asia. During each interglacial the ice retreated to high ground, to be replaced on the plains by tundra and woodland. Man spread into Europe from southern Asia or Africa during the second (Mindel-Riss) glaciation, and across northern Asia during the third (Riss-Würm), when conditions in the Arctic were warmer than they are today. He reached North America during the warm spells in the final glacial phase,

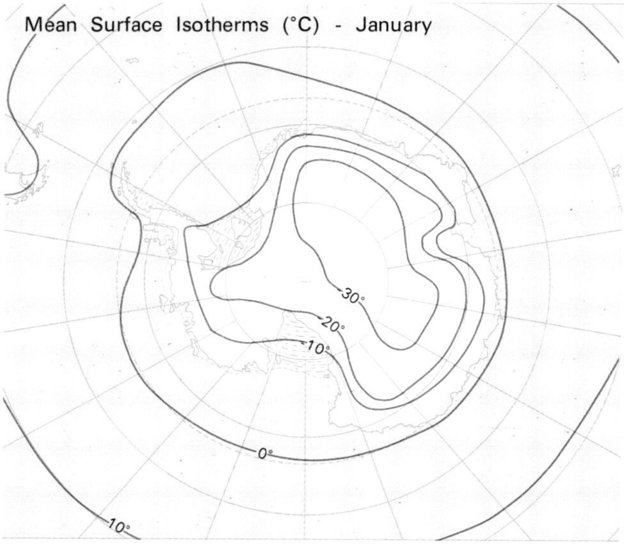

Isotherms in January (above) and July (below). The pole of extreme cold lies over the highest part of the Greater Antarctica ice cap, at a height of over 4,000 metres. Much of the continental coastline is cut by the −2°C. isotherm in summer, the −20°C. isotherm in winter.

Right: the ice-capped mountains of the South Shetland Islands were probably the first Antarctic land to be seen by man, in 1599. Bouvetøya was discovered in 1728, Îles Kerguelen in 1772, and in 1773 Captain James Cook first crossed the Antarctic Circle off Enderby Land. Von Bellingshausen reported an appearance of land close to the Greenwich Meridian in 1820 and in the same year a British merchantman was probably the first to sail within sight of the tip of Antarctic Peninsula. The continent was finally outlined by British, French and U.S. expeditions of the period 1839–43. Details were added by whaling and scientific expeditions between 1895 (when man first set foot on the continent, at Cape Adare) and 1942; since World War II massive land and air surveys have completed the map.

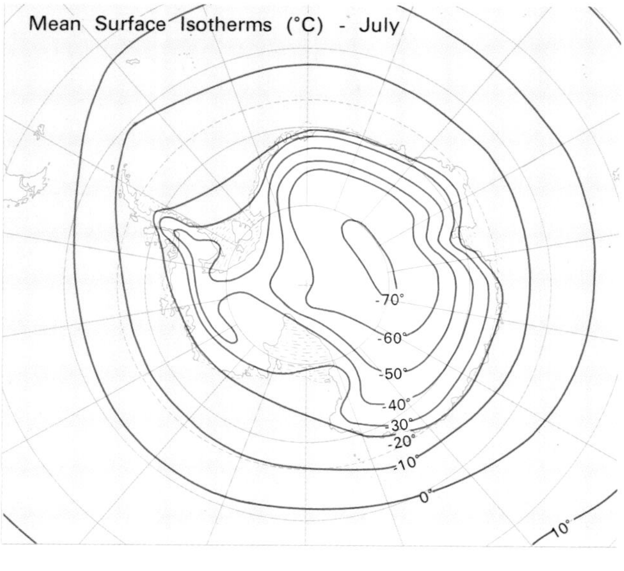

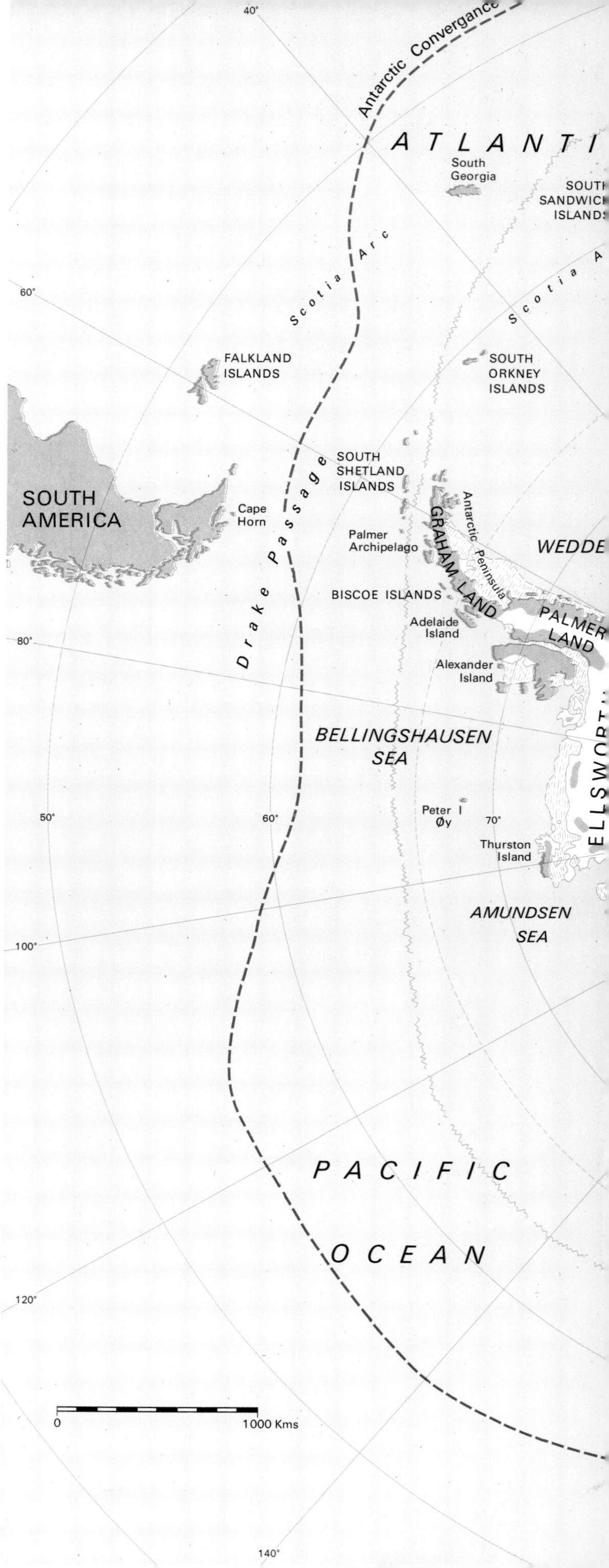

ern Limit of Pack Ice
Bouvetøya
PRINCE EDWARD ISLANDS
Marion I.
OCEAN
ILES CROZET
Antarctic Circle
SØR RONDANE MTS.
DRONNING MAUD LAND
ENDERBY LAND
ILES KERGUELEN
Halley Bay Station
COATS LAND
McDONALD ISLANDS
Heard Island
Berkner I.
Amery Ice Shelf
Greater Antarctica
PENSACOLA MTS.
AMERICAN HIGHLAND
INDIAN OCEAN
WORTH MTS.
SOUTH POLE
Amundsen-Scott Station
WILHELM II LAND
Gaussberg
Queen Mary Coast
TRANSANTARCTIC MOUNTAINS
HORLICK MTS.
Vostok Station
esser arctica
QUEEN MAUD MTS.
RD LAND
Ross Ice Shelf
WINDMILL ISLANDS
WILKES LAND
FORD RANGES
Roosevelt I.
Scott Base
Ross Island
VICTORIA LAND
Sabrina Coast
ROSS SEA
Franklin I.
Mt. Melbourne
Coulman I.
C. Hallett
C. Adare
George V Coast
Adélie Coast
Pte. Géologie
C. Denison
BALLENY ISLANDS
Scott Island
Macquarie Island
West of Greenwich 180° East of Greenwich
0°
20°
40°
60°
80°
100°
120°
140°
160°
50°
60°
70°
80°
AFRICA
ATLANTIC OCEAN
MADAGASCAR
SOUTH AMERICA
INDIAN OCEAN
ANTARCTICA
PACIFIC OCEAN
NEW ZEALAND
AUSTRALIA

A small freshwater pond among the bare rocks of Granite Harbour, Victoria Land. Fed by melt water from winter snowbanks, these small pools thaw completely and reach temperatures of 10°–15°C in summer. Often they contain lively populations of single-celled algae which turn the water green.

Adélie penguins nesting on a raised beach. Formed by uplift of the sea bed since the ice cap retreated, these beaches provide penguin nesting grounds on all the rocky coasts of the continent and islands.

that is, within the last hundred thousand years.

These fluctuations of the northern ice sheets are believed to have resulted from small changes in the amount of solar radiation reaching the earth from the sun. The changes are thought to be due to the combined effects of varying *obliquity* or tilt of the earth's axis, and the earth's *precession in orbit*—which effectively varies the distance between earth and sun, bringing intensified radiation to each hemisphere in turn. Though both factors varied throughout the long ages before the Pleistocene glaciations began, their effects would then have been small. Once cooling had brought the world to the threshold of the ice age, even small changes of intensity would probably have been sufficient to trigger massive changes in the distribution of ice on land. Both obliquity and precession in orbit vary cyclically. Their combined effects over a period of 600,000 years have been calculated by the Hungarian physicist M. Milankovich. There is a most striking correlation between cycles of radiation maxima and minima, calculated by Milankovich, and cyclical fluctuations in climate during the same period, deduced from independent evidence by other workers. This is taken by many to indicate that the climatic cycles of the ice age, and thus the changes in extent of northern glaciation, were controlled by changes in radiation intensity.

The Antarctic ice cap has always been a more stable affair than the northern ice sheets. Once the ice had reached the edges of the continent on every side, it could not extend further without breaking off and floating away. Its massive bulk would probably have protected it against small climatic fluctuations. So its ice sheets do not seem to have advanced and retreated on a continental scale, like those of the northern hemisphere. However, the ice cap has not remained static throughout its long history. Mountain peaks in coastal areas, which now rise far above the surface of the ice, carry moraines and glacial scratches on their shoulders, showing that they were buried deeper in the past. Offshore islands, now isolated, show evidence of having once been linked by ice sheets to the mainland. Much of the continent is ringed by a bank of moraine, between 100 and 300 kilometres offshore and in water up to 500 metres deep, which marks a former edge of the ice mantle. The continental shore bears scars which show that glaciers and ice fronts have advanced or retreated in the past, some within recent years, others many hundreds or thousands of years ago.

These changes are unlikely to have been caused directly by small shifts of climate. According to an old theory, first put forward by A. Penck in 1928 and recently revised by Soviet and U.S. glaciolo-

JANETSCHEK

Moss carpet on a moraine overlooking McMurdo Sound. Such patches of vegetation are the home of Antarctica's richest communities of microscopic animals, "bryosystem" communities.

Frost shattered rocks on a bare hillside support only small patches of moss and lichen with few microscopic animals—"chalikosystem" communities.

JANETSCHEK

gists, they were caused by world-wide changes in sea level resulting from the northern hemisphere glaciations. When the northern ice sheets spread across the plains of America and Eurasia, the water forming them was withdrawn from the sea, and sea level fell. During the most extensive glaciation, sea level the world over fell some 150 metres below its present level. An American geologist, J. T. Hollin, has calculated that such a fall would provide a ring of new coastal land, averaging 90 kilometres in width, all round the continent. As ice built up on the new land, the profile of the ice cap would shift seaward, burying coastal mountains deeper and completely engulfing cliffs, headlands and offshore islands (page 23). During northern interglacial phases, water would return to the ocean. Much of the new coastal ice would find itself afloat, and would break off and sail away under the pressure of offshore winds, restoring the ice mantle to its former shape. Glaciologists working in the McMurdo Sound area have found evidence of four such local glacial advances, which may eventually be correlated in some degree with the four glacial phases of the northern hemisphere.

The most recent retreat of the northern ice sheets began some 20,000 years ago, leaving vast tracts of sterile glacial debris spread across the face of northern Europe, Asia and North America. During the ensuing centuries these have gradually been clothed in soil and vegetation—the grasses and trees of tundra and taiga which now cover Arctic and Subarctic lands. Antarctica, by contrast, remains locked in its ice age. Though it, too, seems to have emerged recently from a more intense spell of glaciation, only a very small fraction of its lowlands have been uncovered. The Soviet geographer Suyetova has calculated that, of the total coastline of 30,000 kilometres, 9% is still formed by ice cliffs, backed by floating shelf ice (45·5%), active glacier (9·5%) or static ice resting on land. In exposed "oasis" regions (page 12) soils remain primitive; even in the most favoured areas of the continent, few patches of vegetation are more than two or three metres across, and the predominant vegetation is mosses and lichens. Antarctica itself has no

MANNERING

Springtails: minute insects from southern Victoria Land, natural size about 2 millimetres. These flightless insects live in rocky crevices and moss clumps, feeding on decaying plants and fungi.

SPAULL

Antarctica's only grass, *Deschampsia antarctica*, grows on the Antarctic Peninsula and the maritime islands, but has not yet been found on the continent. It grows in association with mosses, its roots helping to form organic brown soils.

Usnea, a branching green lichen which grows on sheltered surfaces throughout Antarctica. On the continent it grows sparsely on high inland peaks; on the Peninsula and maritime islands it forms dense mats.

BAS

trees or shrubs, and only two species of flowering plants which grow only on the Peninsula. In comparable northern latitudes, Greenland alone has 600 species of ferns and flowering plants, including both shrubs and trees.

The Antarctic ice cap at present weighs about 27 million billion tonnes, and contains about nine-tenths of the world's ice. Each year it receives some 2,000 billion tonnes of snow and ice, mostly in the form of falling snow or ice spicules falling in coastal regions of Greater Antarctica and on the northern flanks of Lesser Antarctica. A similar amount is lost from the continent each year as evaporation, melt water, snow blown out to sea or, most important of all, as flotillas of huge tabular icebergs. The world's largest bergs, up to 100 kilometres long and towering 40 metres from the water, calve from Antarctica's coastal ice shelves and drift away to sea.

Under the tremendous weight of the ice cap, the underlying continent has sunk about 1,000 metres into the earth's crust. During the maximum glaciation it sank even deeper, and has since emerged slightly with the loss of some of its coastal ice. Evidence for this emergence can be seen on rocky coastal sections of the continent, where raised beaches show the existence of former strand lines. Many millions of Antarctica's smaller penguins nest on raised beaches which must have emerged from the sea during the past few thousand years—certainly since the last retreat of the ice cap edge.

The huge southern polar ice cap chills the whole of the southern hemisphere. South Georgia, a southern island in a latitude equivalent to Britain, is heavily ice-capped; the South Orkney and South Shetland Islands, no further from the equator than their namesakes in the northern hemisphere, are true Antarctic islands (page 24). New Zealand, southern Australia and South America often feel the cold breath of Antarctica. Even the northern hemisphere receives a share of Antarctic cold, for cold submarine currents creep northward from Antarctica to chill the waters of the northern oceans (page 36).

It has been suggested that the world would be better without the south polar ice cap, and that man should use his technical abilities to get rid of it. It may well lie within his power to do so; indeed, whether he likes it or not, he may already have started. He has increased the insulating properties of the atmosphere by releasing into it massive quantities of carbon dioxide and other pollutants, and is warming the earth and atmosphere by vast expenditure of fossil and atomic fuels. Man's activi-

Summer thaw water cannot soak into the frozen ground. Instead it forms surface torrents which cut tunnels through permanent ice, emerging as seasonal rivers which flow strongly toward the sea. When the thaw ends in late summer only the empty water course remains (below). Dronning Maud Land.

ties now contribute significantly to the heat budget of the earth, and may have begun a warming process which cannot be stopped. Though we might enjoy the equable climate of a world without ice caps, we would have to take to the hills. When the Antarctic ice cap melts, sea level the world over will rise between 50 and 100 metres, displacing the half of the world's population who now live on coasts and lowlands.

Though fascinating to glaciologists, the ice cap is frustrating to geologists, who want to study the rocks beneath it, and a matter for regret to biologists. During its spread in the late Pliocene and Pleistocene, it bulldozed what must have been an intriguing flora and fauna from the face of the continent. There is no doubt that, up to the mid-Pliocene, Antarctica like every other continent was fully endowed with plants and animals. Lesser Antarctica may have been joined to South America through the Scotia Arc. Greater Antarctica was certainly linked to Australia up to about 50 million years ago, and must have carried away some of the ancestors of Australasia's fascinating fishes, reptiles, birds, monotreme and marsupial mammals, and plants. Nearly all evidence of this Tertiary flora and fauna has disappeared. The only refugia—areas which missed being glaciated and still retain their original biota—are possibly some of the mountain peaks, whose few handfuls of algae, lichens, mosses and minute animals may be the direct descendants of pre-glacial Antarctic stock. The freshwater streams and lakes may also contain aboriginal flora and fauna, creatures which once lived exclusively among the high peaks of a temperate continent, but spread with the glaciers to inherit the lowlands.

Soils, climate and life

Antarctica has a cold, dry and windy climate. These three qualities between them alienate practically every form of life, and Antarctica in consequence is a desert continent. The harsh climate affects living organisms directly, by freezing them, drying them, and blowing them from their anchorage. Even more discouraging, it constantly erodes or weathers the substrates on which they are attempting to live, and prevents the formation of mature soils.

Soils begin to form when chemical and mechanical erosion break down the surface of rock into gravel, sand, silt and clay. Sorted and consolidated by wind and water, the fine material becomes organized in layers, forming a primitive *ahumic* or *mineral soil.* If this is firm enough to hold moisture between its particles, it attracts a flora of algae and bacteria. In temperate regions where the climate is benign, other plants move in to colonize these damp patches, growing rapidly and adding humus to the soil. *Organic* or *humic soils* are more than a mixture of rock and plant debris. They develop a complex chemistry of their own, releasing minerals from the rocks, buffering excessive acidity or alkalinity, and gradually maturing to provide a hospitable, nutritive substrate for more complex plants.

Polar conditions inhibit soil formation in several ways. Constant cycling of temperature between freezing and thawing disrupts and shatters the rocks. Wind-borne sea salts erode them, and sand and drifting snow abrade their surface. Rock debris is produced faster than plants can colonize it; the resulting immature soils remain uncovered by vegetation and exposed to every change of weather. Constantly stirred by frost action, baked by the sun, washed out by surface streams and blown by the wind, they have no chance of stabilizing. Not surprisingly, polar soils tend to remain ahumic and poor, and to attract little vegetation or animal life. Many thousands of square metres of Antarctica's mineral soils seem to be devoid of all life. Many inland peaks give rise to mineral soils which are virtually sterile, and very dry coastal areas show little evidence even of microscopic life among their stones and gravel. Warmer, damper coastal regions are, however, more hospitable. Some of the best Antarctic soils are found on the Peninsula coast, and in other sectors where snowfall is heavy. Precipitation washes harmful concentrations of minerals down through the soil, and forms a shallow underground reservoir of moisture to keep plants from drying out in the summer. Where it occurs, thick mats of algae, lichens and mosses form, with a busy fauna of mites and insects living among them. But even the most favourable sites have failed to produce mature organic soils. Though peat-like deposits several centimetres thick are recorded as far south as McMurdo Sound, and richly organic *ornithogenic soils* occur wherever penguins congregate to breed (page 88), no true *brown soils* like those of the Arctic tundra have so far been reported anywhere on the continent.

With harsh climate and raw, uncongenial soils, it is not surprising to find that Antarctica is poor in plant and animal species. The most prominent plants are algae (about 360 terrestrial and freshwater species) and lichens (about 400 species). About 70 species of mosses are listed, but there are no ferns and only two species of flowering plants—a grass, *Deschampsia antarctica* and a pink, *Colobanthus quitensis*—both restricted to the warmer maritime region of Antarctic Peninsula. All grow very slowly; few are more than three centimetres tall, and only in especially favoured areas is the ground even partly covered with vegetation. There is no forest or taiga, no muskeg or rich swampland—not a single woody shrub or herb on the whole continent. The sparsity and slow growth of vegetation help to explain why there are no large herbivorous mammals in Antarctica—no reindeer, musk oxen, hares, voles, lemmings or ptarmigan to match those of the far north. Antarctic herbivores do exist, but they are tiny insects and mites (page 28), none bigger than a mosquito, which feed mainly on algae, fungi, and the broken-down remains of plant material. Without large herbivores to feed on, there can be no terrestrial carnivores—no wolves, foxes, weasels, coyotes or polar bears. Antarctica's largest terrestrial carnivores are mites about one millimetre across, which live in the debris among the vegetation and feed on herbivorous insects.

Nor is there a native human population. It is not impossible that Eskimo-like people—and perhaps other large predatory animals—could have survived in favoured coastal regions of Antarctica where birds and seals are plentiful. But the question has never arisen; with the spread of the ice cap Antarctica lost all the large land mammals it possessed, and throughout the period of human history it has been isolated from the rest of the world by at least 1,000 kilometres of storm-ridden ocean.

How cold is Antarctica? The southern ice cap in winter is by far the coldest place on earth. Vostok, a Soviet research station 3,488 metres up on the ice dome of Greater Antarctica, lies close to the pole of extreme cold. At depths of winter in August, its mean daily air temperature is −68·4°C., its highest temperature (recorded over a nine-year period) −44·9°C., and its lowest −88·3°C. At the height of summer in December, mean temperature rises to −32·7°C.; the highest air temperature ever recorded at Vostok is −21·0°C. The South Pole, slightly lower on the plateau, is correspondingly warmer, with a July mean of −59·2°C., extreme winter low of −80·6°C., and December mean of −28·1°C. (page 60). Siberia and the northern prairies of Manitoba and Saskatchewan, the coldest regions of the north in winter, are closer to sea level and never experience such intense cold. Their

BURNIP

Polar greenhouse. Lichens growing in a damp crevice are protected from winds and intense cold by a thin shelter of ice.

lowest mean and extreme winter temperatures are some 20°C. higher than those of the south polar plateau. Their summers are very much warmer, with temperatures well above freezing point for days or weeks on end.

No plants, and no animals but man, are rash enough to live in the extreme low temperatures of the high Antarctic plateau. The nearest living organisms are the sparse mosses, lichens and algae occurring at heights of almost 2,000 metres among the peaks of the Horlick and Queen Maud Mountains, only 260 kilometres from the South Pole. Their winters may well be as cold and sunless as those of the Pole itself, but summers are very much warmer among the heat-absorbing rocks. Not only the rocks, but the lichens themselves absorb radiation and warm up in the sun. Insects and mites have been found in very similar situations within 600 kilometres of the Pole. On sunny December days temperatures within and about the lichens rise to 6° or 7°C., providing a sheltered micro-environment very much warmer than the surrounding air.

The Antarctic coast is warmer and more hospitable than the interior. Hardy folk who have wintered at plateau stations refer to the coast as the "banana belt", and shed half their warm clothing when they

Cape Denison. The huts of Sir Douglas Mawson's expedition (1912–1914), now almost completely snow-covered. Receiving constant katabatic winds from the high interior, this corner of Antarctica is one of the windiest on earth.

PRÉVOST

return to the coast on their way home. This is not entirely bravado. Even in far southern sections of the coast—McMurdo Sound, or the southern Weddell Sea—summers are tempered by the nearness of open sea, and around midday air temperatures often rise two or three degrees above freezing point. They seldom rise more, for there is usually enough snow about to absorb surplus heat and keep both air and ground close to freezing point. When ice covers the sea in winter (page 39) it cuts off warmth from the coast. Southern coastal stations especially suffer long continental winters (McMurdo, Ellsworth: page 24). Those further north remain free of sea ice longer in spring and autumn, and their winters are never so cold. By far the warmest region of the continent is the western flank of Antarctic Peninsula, which lies squarely across the track of westerly depressions. Winter and summer alike, it receives warm, snow-laden winds from the southern Pacific Ocean. Both coast and offlying islands are heavily snow-capped, and the damp summer atmosphere, with long spells of above-freezing temperatures, encourages a relatively luxuriant growth of ground vegetation.

Cold is not the main problem of polar organisms. In most areas drought, rather than cold, sets a limit to plant and animal life. Much of central Antarctica receives the equivalent of only 15 to 30 centimetres of rainfall each year. Coastal regions usually have more, but much of the snow evaporates or blows out to sea, and never becomes available for living organisms. Mountains, porous moraines and shingle beaches are especially dry, for the meagre snows of winter melt quickly in spring. Except in a few warm, sheltered places along Antarctic Peninsula, there are no mats of organic soil or vegetation to hold the moisture. For lichens growing on rock faces, free water may be available for only a few days or hours each year. Many snow-free areas of the Antarctic coastline, where ground temperatures in summer reach 15° to 20°C., remain deserts because there is not enough ground water to support either insect or plant life.

Snow banks often provide good living conditions for plants. Throughout winter they blanket the ground and its inhabitants, protecting them from the fiercest cold of July and August, and from the erosion of blown sand and drifting snow. In September and October they start to thaw, and measure out their stored water slowly through spring and summer. As the snow level falls the sun penetrates more deeply. The plants beneath find themselves in a warm, moist greenhouse, living in shelter at temperatures consistently higher than those outside in the open.

LEWIS SMITH

Strong winds deposit snowdrifts over nesting Adélie penguins. Incubating birds usually survive but snow-laden winds are extremely hazardous for chicks.

Antarctica is the coldest continent, and probably also the windiest. The combination of intense cold and wind, devastating to any warm-blooded animal which is trying to keep warm, has given Antarctica its reputation for danger. Winds on the high plateau are mostly light, tending to blow downhill along the gentle slopes of the ice dome. At the edges of the plateau they accelerate, lifting and blowing clouds of snow high into the air. The strongest winds are felt down the long coastal slopes, especially those of Greater Antarctica. In 1912 Douglas Mawson, the Australian explorer, built his base at Cape Denison, on the Adélie Coast of Wilkes Land. After a few months' residence it became clear that he had picked the windiest spot in the world to work in. Average wind speeds for March, April and May of his first year were respectively 22, 23·2 and 27·3 metres per second. The day's mean wind speed on 11 May was 36 metres per second and on 19 May, 40 metres per second. Gusts of over 67 metres per second were commonplace, and the average wind speed for the two years' stay was 20 metres per second. Practically all of these were katabatic (downslope) winds, triggered by cyclonic distur-

Pack ice off Marble Point, McMurdo Sound. Here the land ice (left) has retreated to expose bare, rocky shore with a fringe of raised beaches.

bances and helped on their way by the long, shallow slopes of the coastal ice sheets.

Cape Denison was remarkable for the constancy of its winds. Equally strong winds, though fortunately less persistent ones, occur everywhere in Antarctica from time to time, notably when depressions—gyrating masses of warm and cold air—sweep in from the sea to disturb local patterns of circulation. As a warm-blooded animal, man is especially sensitive to cold moving air. Standing tall above the ground, he meets its full impact and may lose a great amount of heat to it. Wind can blind, buffet and disorientate him, and bring him close to the limits of his endurance far quicker than cold still air. Other animals probably feel it less. Flying birds move with it. Penguins and seals, living closer than man to the ground, more readily find wind-breaks to protect them, and often sleep comfortably as drifting snow builds a protective blanket around them. Insects and mites, safe among the foliage of their vegetation, may never notice the wind at all. For plants, strong winds are damaging and drying agents, but the tough, leathery skins of many lichens are an important protective barrier against the boisterous attentions of such winds. New organisms—spiders, moths, seeds, spores, living fragments of plants—are constantly being carried to Antarctica from South America and elsewhere, and wind transport probably accounts for the spread of plants and animals about the continent.

The Antarctic Ocean

MICKLEBURGH

Sail south from Montevideo, Cape Town or Sydney, and you enter the Southern Ocean, the broad band of turbulent water which rings the southern hemisphere between latitude 40°S. and Antarctica. Head south-southwest, for westerly gales and a strong surface current will carry you eastward across the forties and fifties. Stay on deck if you can, and marvel at the daily air pageant laid on by the birds. Nowhere in the world will you see greater congregations of sea birds, more spectacular displays of flying skill, than among the albatrosses, mollymawks, fulmars, prions and other petrels of the Southern Ocean.

It is never warm at sea in the forties and fifties, even at the height of summer. Wind, rain and spray, always a little colder than yesterday's, keep you awake on watch. But somewhere between 49°S. and 55°S. there is a sharp drop in temperature, and a subtle change in ocean and atmosphere. If the weather is calm and the sea smooth, you see a fog bank, a zone of turbulence and foam in the water, a concentration of birds squabbling and fluttering in flocks on the sea. The engine room reports a drop in sea temperature of two or three degrees, in summer from 5° or 6°C. to 3°C. The air has a new, sharper bite. This is the *Antarctic Convergence*, one of the great

Previous page: HMS *Endurance* in the Southern Ocean.

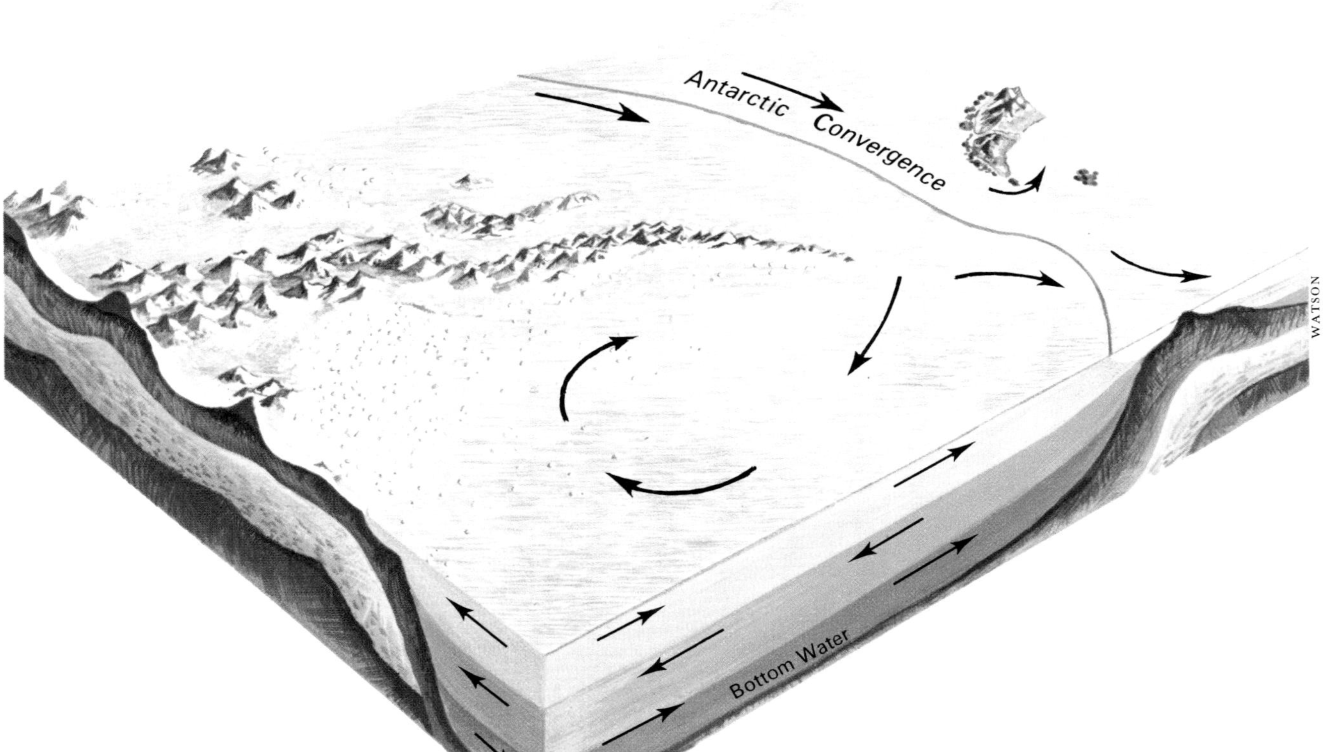

Antarctic water masses. A thin layer of *Antarctic Surface Water* is blown north and east by the broad band of westerly winds which encircles the earth in high southern latitudes. This water is pulled up from below at the Antarctic Divergence, a narrow region of up-welling close to the continental coast. In the Weddell Sea area (shown here) very cold water in contact with the continent sinks under its own weight, to form a northward-flowing deep current—*Antarctic Bottom Water*. The two north-flowing water masses are balanced by the south-flowing *Warm Deep Current*. At the Antarctic Convergence Antarctic and Subantarctic surface waters meet.

USNS *Eltanin* at the Antarctic Convergence.

natural boundaries in the surface waters of the ocean. You are crossing from the warmer Subantarctic to the cold Antarctic zone of the Southern Ocean—to the ring of water surrounding the continent, which we are calling the Antarctic Ocean.

Sea surface and air temperatures fall steadily as you head south. Clouds thicken; the sky is iron grey and the sea black. Rain turns to sleet, then to wet, blinding snow. Snowflakes become mush on the decks and clog the clear-view screens of the wheelhouse. Birds are still plentiful. Great wandering albatrosses with three-metre wing span swoop over at mast height, settling in the wake to squabble over galley refuse. Flurries of prions, like blue-grey snowflakes, wheel in and out of the mist; storm petrels patter the wavetops alongside the ship. The radar hums busily, searching for icebergs. Sea temperature falls to zero, then to —1·0°C. and lower, for this is salt water, freezing about —1·9°C. As sea temperature drops, the horizon ahead brightens. There is a sharp rap against the bow and a rattle alongside; the ship's speed slackens as the first floating ice appears. Now there is a vivid whiteness everywhere—glare from the snow-covered ice floes reflecting in snow-white cloud, and inky lanes of ultramarine or black water. Further south the sky clears and the sun dazzles. Grey Antarctic petrels, chequered pintado petrels and white snow petrels with black legs and beady black eyes flutter about the wake. This is the world of the pack ice, a clean, calm, and astonishingly quiet world after the bustle and commotion of the open sea.

The world of the pack ice. Fragments of the winter ice sheet, and tabular icebergs drifting with wind and current off the coast of Antarctica.

MICKLEBURGH

Antarctic water masses

The waters of the Antarctic Ocean are deep and layered like a sandwich. You cannot reach the continent without crossing water more than 3,000 metres deep: some of it is over 5,000 metres deep, and the deepest trough, east of the Scotia Arc, descends over 8,000 metres. The layers of the sandwich are water masses of differing temperature and salinity, which flow constantly eastward about Antarctica under the steady coercion of the west winds. The middle layer spirals south as well as east, carrying nutrients and a touch of warmth from temperate and tropical regions. The upper and lower layers move gradually northward, carrying cold away from Antarctica into the tropics and beyond (page 36).

The uppermost layer of water, and the one which most directly affects Antarctic animals and plants, is a thin veneer between 70 and 200 metres deep, called *Antarctic Surface Water*. This originates at the *Antarctic Divergence,* a narrow zone in the southern Antarctic Ocean where subsurface water is pulled upward between two divergent bands of surface water. The divergence is caused by a sharp difference in prevailing winds. South of the zone, prevailing easterlies with an onshore component pull surface waters westward about the continent. North of the zone, the strongest and most persistent winds are westerlies. They are responsible for pushing the main body of Antarctic Surface Water slightly north of east, and for pulling replacement water up from below at the Divergence.

Antarctic Surface Water is chilled by contact with ice and cold air from the continent, and its temperature remains below $-1°C.$ throughout winter. In summer its northern boundary warms slightly, reaching temperatures of 3·5° to 4°C. at the Convergence. Melt water from the continent and from melting floes, bergs and snow reduces its salinity in summer to the low value of 34·1 parts per thousand (see below for comparisons). This combination of low temperature and low salinity distinguishes Antarctic Surface Water, and helps us to trace its remarkable wanderings in the other oceans of the world.

At the Antarctic Convergence, Antarctic Surface Water moving eastward and slightly north comes into contact with a mass of warmer and rather more saline water—*Subantarctic Surface Water.* Forming a broad, continuous belt about the Southern Ocean, Subantarctic Surface Water is generally some 3°C. warmer than Antarctic Surface Water throughout the year, with salinities of 34 to 35 parts per thousand. It, too, moves eastward, but with a slightly southerly component. Where the two water masses run against each other, the denser Antarctic water sinks below its neighbour and continues its slow, spiralling journey under a new name—the *Antarctic Intermediate Current.* There is very little mixing between the two water masses at the Convergence. Though the cold, dilute water increases in temperature and salinity as it moves northward between warmer, saltier water masses, it is still detectable as a subsurface current north of the tropics in the Atlantic Ocean. It underlies the surface of the Indian and Pacific Oceans, washing upward against submarine shelves to cool the coastal waters of New Zealand, southern Australia, and many oceanic islands.

The sudden mixing which occurs at the Convergence is enough to kill or incapacitate many tiny animals and plants of the plankton (page 44) which cannot tolerate rapid changes in temperature or salinity of their environment. So the Convergence zone is popular among sea birds, which gorge themselves on the dead and dying animals of the zooplankton. In the Atlantic sector of the Southern Ocean the Convergence is often well defined and at times clearly visible. In the Pacific sector it may be less evident, though it can usually be detected by the drop in temperature of about 3°C. which distinguishes the two water masses. Its position varies only slightly from season to season, year to year, or even from century to century.

The Convergence is a good biogeographical

MICKLEBURGH

Sea ice: the first stages. Calm autumn evenings with clear skies allow the surface layer of the sea to chill below freezing point, forming mats of ice crystals.

Pancake ice. Slight swell causes new sea ice to break into pans which rub together and develop raised edges. Further cooling cements the pancakes together.

BONE

boundary, for the two water masses support different species of planktonic plants and animals, different fishes, and even different species or subspecies of birds. The separation is not complete, but there is a marked change of flora and fauna on crossing from one water mass to the other. Even the ocean floor deposits reflect the position of the Convergence far above them. Between continent and Convergence the sea bed is covered with diatomaceous ooze, a yellowish mud composed of fine rock flour and glacial grit, heavily impregnated with the silica shells of diatoms. These are the tiny plants of the plankton, and their frustules or minute outer cases, almost insoluble, snow constantly down upon the sea bed from Antarctic Surface Waters above. North of the Convergence, beneath warmer Subantarctic Surface Water, siliceous diatoms are replaced in the ooze by calcareous shells of globigerina—minute protozoans which float in the surface waters and, like diatoms, fall to the sea bed when they die. As deep-water dredging has shown, the floor of the ocean is liberally scattered with sponges, molluscs and other filter-feeding animals, which live on the bounty from above.

Close to the shore of the continent, especially in the shallow Weddell Sea, extremely cold conditions favour the formation of a second northward-flowing water mass—*Antarctic Bottom Water.* This is cold water, generally with temperature below −0·5°C. Because sea ice has formed within it, withdrawing water but leaving salt behind (page 40), it is also highly saline with 34·5 or more parts of salt per thousand. High density causes it to sink close inshore, whence it creeps northward and eastward to form a continuous ring about the continent. Like Antarctic Surface Water it spreads far north into the Atlantic and Pacific Oceans, carrying southern polar cold into the northern hemisphere.

The third Antarctic water mass is the southward-flowing *Warm Deep Current.* This is a layer up to 2,000 metres thick, which originates as surface waters of the Atlantic and possibly of the Indian and Pacific Oceans. With salinities of 34·5 to 34·7 parts per thousand and temperatures between 1° and 3°C., the Warm Deep Current can be traced in its southern journey beneath Subantarctic and Antarctic Surface Waters. It reaches the surface along the broad boundary or front called the Antarctic Divergence, welling up to become Antarctic Surface Water. The concentrated nutrients which it brings to southern waters yield a rich return in plant and animal life.

Sea ice

Each winter the sea about Antarctica freezes to a depth of one to three metres, and to a distance of 100 to 200 kilometres offshore. On clear nights in late autumn surface waters chill to freezing point and a slurry of ice crystals appears. If the sea is calm, the crystals join and harden, forming fibrous *young ice.* A gentle swell breaks the thin layer into *pancake ice*—small rafts of ice with edges raised due to rubbing on each other. As more crystals form, the pancakes freeze together again, damping further movement and becoming a solid layer several

Ice flowers—bundles of crystals which form when mist settles on new sea ice. Their "petals" are fine crystalline sheets and needles of ice, sometimes over 10 centimetres long.

BONE

Once formed, the ice sheet usually remains fast to the land through the winter. Cormorants and other birds may have to travel far to find open water for feeding. South Orkney Islands.

In late spring fast ice breaks up and floats out to sea, becoming pack ice. Heavy snowdrifts which form along the shore break out with it. The Adélie penguins are finding a safe path to the water.

centimetres thick. Two or three days and nights of calm weather and sub-zero temperatures will thicken the ice further. Now it has become *fast ice*, so called because it is fast to the land, and it may remain fast, thickening steadily throughout the winter, until late in the following spring. Ice formed in this way is frozen fresh water. Concentrated brine remains in the cavities between its crystals, slowly leaching out to leave surface waters slightly more salty than before.

Fast ice forms all around the continent, except in *polynyas*—areas where strong offshore winds or currents keep the sea open even in winter. It moves up and down with the tide, forming a complex hinge or tide crack where it joins onto the land. Cracks or leads also appear due to pressures and tensions within the ice sheet, especially among islands and stranded icebergs, and off the moving faces of glaciers and active shelf ice. They are very important in the life of birds and seals, allowing birds to feed in open water along the leads, and seals to breathe and emerge from the sea to lie on the ice.

Winter snows build up on the fast ice, their weight forcing it down below sea level and allowing salt water to float over it. Dense layers of ice platelets form underneath, adding substantially to the total thickness and making a delightful crystal-palace habitat for diatoms and surface-feeding fishes (page 45). After a hard winter and a cold spring, fast ice may remain in position for several seasons. Then it is called *bay ice*; as such it has been known to persist for years on end, eventually building up to form *shelf ice* many metres thick. However, fast ice usually begins to rot at the start of its first summer, and during the summer breaks into large pans or *ice floes* which are driven in masses along the coast by offshore winds. Now it has become *pack ice*. Once at sea it drifts slowly about the continent, joining older pack ice left over from previous years. At first it drifts westward, driven by easterly winds which prevail in a narrow zone a few miles from the continental shore. Further from the coast, beyond latitude 65°S., it meets the southern edge of the westerly wind belt, and begins to circle eastward. Local winds, concentrations of icebergs, islands and coasts build up pressures in the pack ice, forcing floes to override and topple each other. Local gyres of pack ice form in the Weddell and Ross Seas, and heavy concentrations of pack ice—too dense even for the most powerful modern icebreakers to fight—lie off the Pacific coast of Lesser Antarctica.

The outer edge of the pack ice shifts with the seasons. In winter and early spring (August and September) it lies on average some 800 kilometres from the Antarctic coast. This is the *northern limit of pack ice* shown on page 24. It encloses a sea area of nearly 19 square kilometres—almost half as much again as the area of the continent. In October and November the sun warms the ocean surface and the pack ice begins to melt, its outer edge retreating rapidly through December, January and February.

SPAULL

MANNERING

By March only a narrow band of floes remains along most of the Antarctic coast, with a wider band off the Pacific side of Lesser Antarctica and massive semi-permanent gyres of solid pack in the Weddell and eastern Ross Seas. This amounts to no more than $2\frac{1}{2}$ million square kilometres, between one-seventh and one-eighth of the winter area. Local winds and current tend to carry the summer pack ice away from the shore, so that an unprotected ship can often sail without danger in coastal waters as far south as McMurdo Sound (78°S.) in the Ross Sea, and Halley Bay (75°S.) in the Weddell Sea, during February and March.

Floating sea ice affects the ecology of many Antarctic plants and animals. Though translucent when new, it rapidly becomes opaque when thick snow covers it, cutting down the amount of light entering the sea and restricting the growth of both bottom-living and surface-living algae. This may not be vitally important, however; although very little light enters the sea between late autumn and early spring when the sun is low or completely absent, enough light penetrates in summer to promote a substantial growth of algae under and between the broken floes of the pack ice. Floating ice keeps the sea cool in summer, both by reflecting energy back into space and by using up warmth in melting. In winter it keeps the sea warm, by forming an insulating layer on the surface. Though cold, the sea in winter is much warmer than atmosphere or sky; the layer of ice inhibits both the flow of heat from ocean to atmosphere and the radiation of long-wave energy to space.

Because ice blankets the sea, its presence means that the sea cannot warm the coasts. Islands surrounded by ice throughout the year are almost as cold as the shore of Antarctica itself, and their soils, vegetation and microfauna match those of the continent. The South Shetland and South Orkney Islands, surrounded in winter by pack or fast ice, but by loose pack or open water in summer, have long winters of continental severity but much warmer summers. Their soils mature further, and their vegetation, though restricted mainly to lichens and mosses, includes a wider variety of species. Islands like South Georgia and Kerguelen, lying north of the winter limit of pack ice, receive the full benefit of open water throughout the year. Their winters are very much milder, with temperatures seldom dropping far below freezing point, and their summers are long enough to encourage the growth of tussock grasses, small woody perennials, and lush growth of mosses, ferns and other plants.

Sea ice is the year-round home of Weddell, crabeater and Ross seals. Only Weddells are normally found close to land or ashore: crabeater and Ross seals spend their whole lives on and about the pack ice, catching their food in leads and hauling out onto the floes to rest and produce their pups. Leopard seals too, find ice floes a convenient floating base. Emperor penguins breed on sea ice in winter, forming colonies in April as soon as the ice is strong enough to support their weight, and preparing their chicks for independence just as the ice breaks out in summer. Other Antarctic penguins breed on land in summer, but live on the pack ice—where the climate is generally milder and food closer at hand—in autumn, winter and spring. Flying birds roost on the pack ice, and feed in leads and lanes between the floes. Hunting by sight from the air, they probably see food more clearly in the calm water of leads than in the rough, broken waters of the open sea.

However beneficial to animals of deep water, ice does not encourage plants and animals to settle in shallow water or intertidal zones of the shore. Continental shores, and those of islands within the pack ice zone, are virtually barren; the only creatures to be found there are a few tiny worms,

MANNERING

Right: This diagram shows some of the major food chains in Antarctic waters. All energy is derived ultimately from sunlight, which phytoplankton—the minute floating plants of the ocean surface—absorbs and converts to complex food chemicals.

Ice foot: thick ice left along the shore when the fast ice has broken away. The ice foot usually persists throughout the summer, allowing few intertidal plants or animals to colonize the coast.

Pack ice forms a floating platform for penguins and seals, which dive off the floes to feed. It allows them to extend their feeding range far from coastal waters.

CURTSINGER

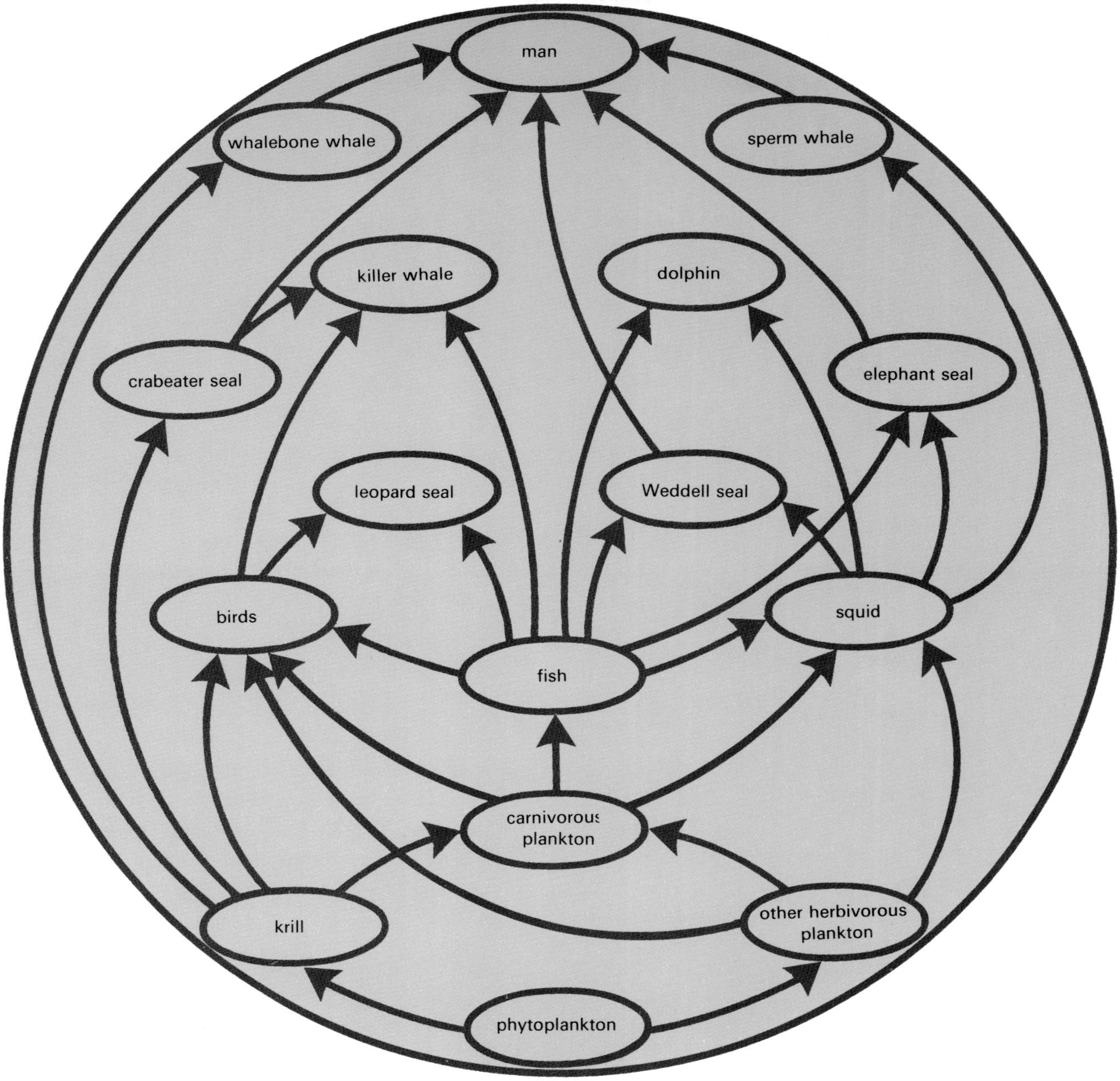

FLOOKES (AFTER HART, EL SAYED)

deeply buried in crevices between rocks. Limpets, barnacles, and seaweeds, the normal inhabitants of warmer coasts, would almost certainly find it impossible to survive with floes grinding against the shore in spring and autumn, and solid fast ice covering them through a long winter. Shallow-water flora and fauna are discouraged by anchor ice, large platelets of ice which form on the sea bed down to depths of about 33 metres. It appears during very cold spells, and grows like white lettuce to form mats up to 40 centimetres deep. Slight currents may lift huge chunks of anchor ice, which float up to the surface carrying sponges, starfishes, algae, and other inhabitants of the sea floor. Not surprisingly, the zone immediately below low-water mark of colder coasts is relatively bare, though seaweeds, marine worms, sponges, starfishes, brittle-stars and a host of other animals flourish lower down on the continental shelf. Islands lying north of the limit of pack ice have relatively rich intertidal and sub-littoral zones, usually with massive banks of kelp growing in the water immediately offshore.

The plankton

All who have sailed the waters surrounding Antarctica are struck by the extraordinary abundance of life, in what at first seems to be a cold, dismal, inhospitable stretch of ocean. Huge flocks of birds fill the sky. Shoals of inquisitive penguins circle the ship, commenting raucously and showing off their swimming and diving skills. Seals lie scattered on the ice floes like over-sized slugs. Killer whales blow among the pack ice; up to the 1950s whalebone whales were prominent in open

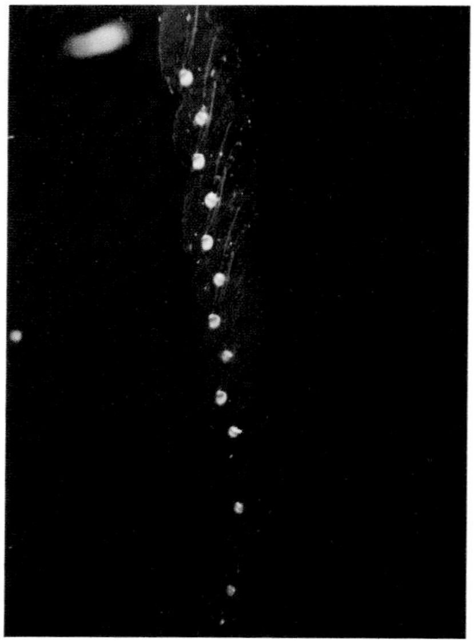

SHABICA

Jelly-like masses of a colonial salp. Transparency helps to make planktonic animals almost invisible to their many predators in the water. Salps form chains 3 to 4 metres long, and are frequently encountered as dense shoals in the Southern Ocean.

US NAVY

Euphausia superba, the common euphausiid crustacean of offshore Antarctic waters. Dense swarms of this creature form the main food of the great whales, crabeater seals and many Antarctic birds. Length 6 to 9 centimetres.

water, steaming lazily in groups among the shoals of crustaceans which were their food. Huge patches of discoloured water, green, brown or blood-red, show the presence of plankton—minute floating plants and animals in concentrations as thick as a clear but interesting broth. This spectacular parade of animal and plant life is a true indication of the biological wealth of the Antarctic Ocean. Acre for acre, Antarctic surface waters are indeed among the world's richest, with overall productivity about four times that of other oceans. The richest sectors in summer yield more vegetable matter than good agricultural land, more carbohydrate, fat and animal protein than well-managed pasture. Some scientists look hopefully toward Antarctic waters as an untapped food reserve for man, a hamper as yet unlocked in the world's emptying larder.

The wealth of southern waters rests partly in their low temperature, partly in their high mineral content. Cold water holds more dissolved carbon dioxide and oxygen in solution than warm water; thus the gases needed by plants for photosynthesis and by animals for respiration are readily available. Then the constant upwelling of mineral-rich water from below brings ample phosphates, nitrates, and other essential compounds to the surface, putting them at the disposal of plants in the well-lighted zone where photosynthesis and growth occur. Finally, the long hours of daylight allow plants to photosynthesize almost continuously throughout summer. The intense cold of polar water restricts the number of species adapted to live in it, but the few which have adapted are present in astronomical numbers.

Phytoplankton Every marine animal, whether sponge, shrimp, albatross or whale, depends ultimately for its livelihood on the phytoplankton, the minute plants of the plankton. In Antarctic waters practically all are diatoms—tiny, single-celled plants, each contained in an elaborately sculptured box of glass-like silica. Individual diatoms are pale brown or green. Massed together they give surface waters their greenish summer tinge, and stain the edges of the ice floes red-brown. Many are shaped like pill boxes or pencils; some form chains or linked strands, with whiskers and bristles which help them to keep up at the surface. Dinoflagellates, another form of unicellular green plant, are also present, but diatoms alone make up ninety-nine per cent of the bulk of every catch.

The minute plants are eaten voraciously by myriads of small, floating animals—the zooplankton—which browse continuously among them. Prominent among these are krill (shrimp-like crustaceans of the family Euphausiidae), other kinds of crustacea and larval fishes. They are accompanied by millions of pelagic molluscs, salps, jelly-fish and arrow-worms, many of which are carnivorous and feed on the herbivores. Zooplankton is the food of birds, fishes, squid, whalebone whales and seals, which unite with it to form a closely integrated community. Some of the more important feeding relationships within the surface community are represented on page 43. Both plants and animals of the plankton sink from surface waters when they die. Their remains, together with part-digested debris and the carcases of larger animals, eventually reach the sea bottom, where they contribute to the welfare of filter-feeding and scavenging animals which abound there.

The standing crop of phytoplankton (i.e. the total amount present in a water mass) can be estimated fairly simply from the amount of photosynthetic green pigment—chlorophyll—dis-

SHABICA

Ctenophore or comb-jelly. Rows of beating hair-like cilia drive this translucent animal—a distant relative of jellyfish—slowly through the water. Over 99 per cent water themselves, comb jellies are a poor catch for predators.

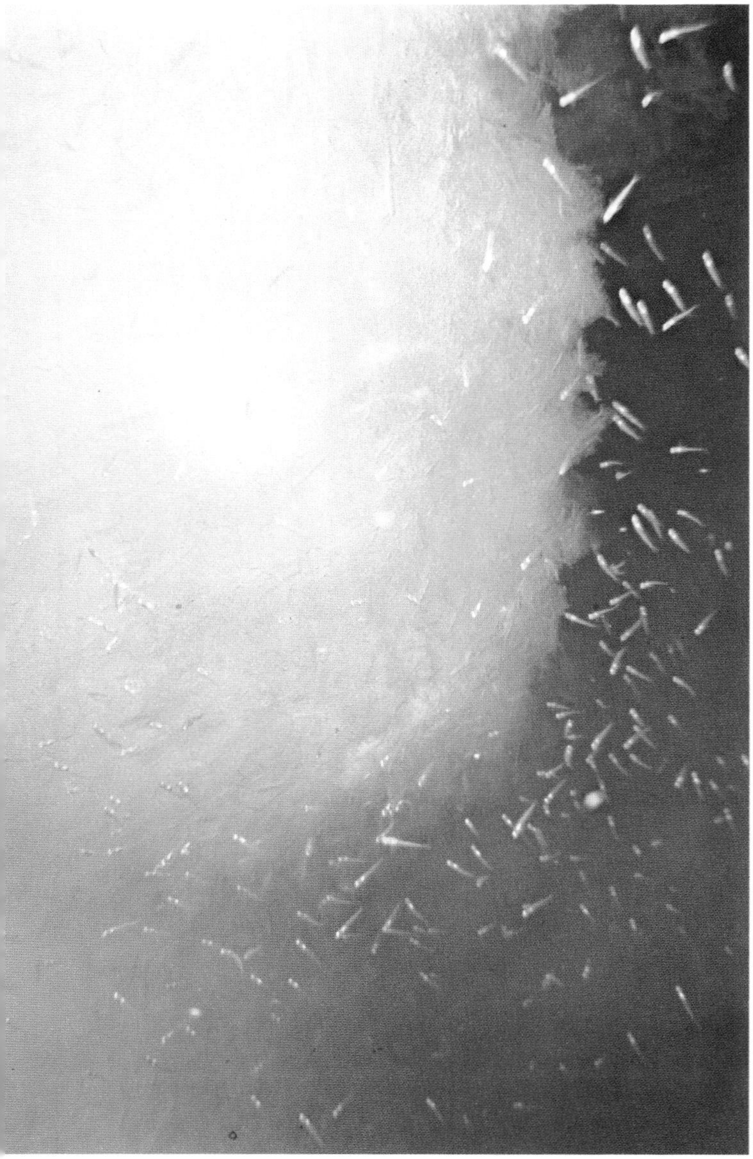

KOOYMAN

SHABICA

Planktonic pteropod, a distant relative of the snail and squid.

Larval fishes feeding on a rich accumulation of diatoms under an ice flow. Many species of bottom-living fishes have planktonic larvae, which feed for a season in surface waters before descending to the sea bed.

Pelagic polychaete—a sea-going relative of the rag-worm and earth-worm. Undulating movements help it to swim slowly through the plankton in search of food.

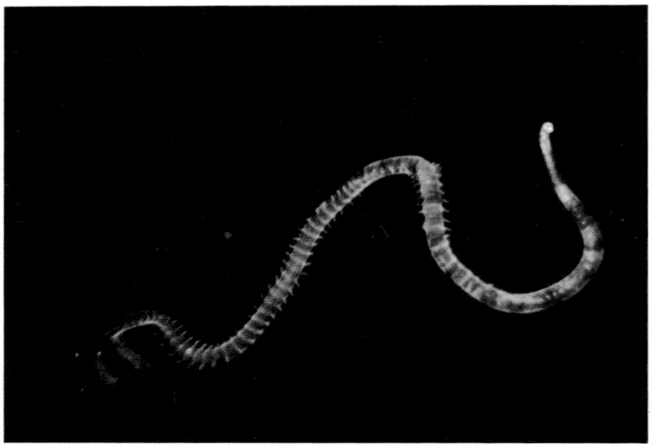

SHABICA

colouring the water. Yield—the productivity of the water—is measured by assessing the amount of radioactive carbon-14 assimilated by a sample of plants representing the population of a known area of ocean surface. From the results of recent U.S., Soviet and Japanese expeditions, both standing crop and yield are known to vary widely across the length and breadth of the Antarctic Ocean. Apart from their very marked seasonal variations, both are highest in waters close to the continent and among islands, where turbulence brings underlying layers to the surface, and lowest in mid-ocean where there is less mixing. According to Sayed Z. El-Sayed, a U.S. biologist who has worked intensively in the far southern ocean, small standing crops of phytoplankton (less than 0·49 milligrams of chlorophyll per cubic metre of surface water) are found close to the Antarctic Convergence, and in the wide open spaces of the western Atlantic and eastern Pacific sectors of the Antarctic Ocean, the Drake Passage which connects them, and the ice-packed Weddell and Bellingshausen Seas which form their southern extremities. Very much larger standing crops (between 1·0 and 10·0 milligrams per cubic metre) occur in the Scotia Sea, south-western Weddell Sea, along the western shore of Antarctic Peninsula, and also in the shallow Ross Sea on the opposite side of the continent. Inshore waters close to Antarctic Peninsula and the Scotia Arc usually contain between five and ten times the bulk of standing crop of neighbouring oceanic waters.

Similarly, rich inshore waters can produce up to five times as much phytoplankton per square metre as the open Antarctic Ocean. This allows

KOOYMAN

KOOYMAN

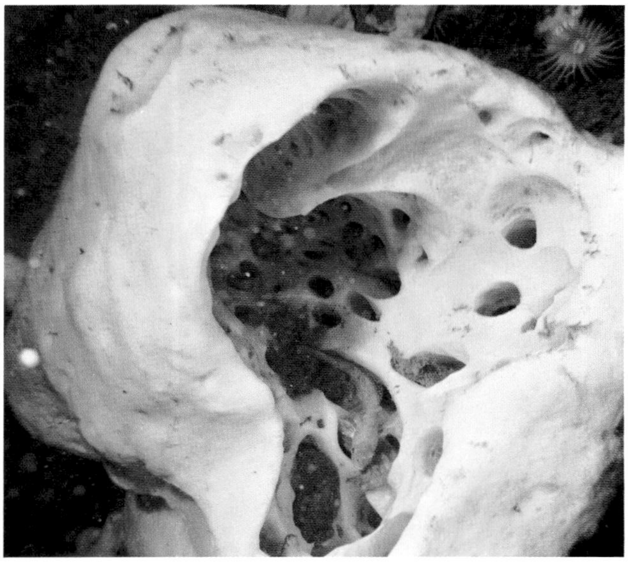
KOOYMAN

Top: A dense sea-bed community of scavengers and detritus feeders from 80–100 metres: a red compound ascidian with sea anemones and starfish, *Odantaster validus*. In the right foreground is a coral (*Alcyonium*).

Centre: Sea anemone *Utricinopsis antarctica* on rocky sea floor at 128 metres. Anemones trap living prey—often small shrimps and fish—among their tentacles, immobilize them with poison darts and digest them in their hollow, sac-like stem.

Left: A small fish, probably a nototheniid, hides in the main exhalent channel of a sponge, 200 metres down on the sea bed. Seldom taken as food, sponges offer lodging to smaller species.

Above, right: An erect octocoral with the common echinoid *Sterechinus* browsing over rocks encrusted with pink algae.

Right: Nudibranch mollusc from McMurdo Sound. Nudibranchs are sea slugs, generally found in shallow water. Like land-bound slugs and snails, they browse on vegetation, rasping the plants themselves and the film of diatoms which grows on their fronds.

KOOYMAN

CURTSINGER

them to support huge stocks of zooplankton, and makes them the main haunts of birds, seals and whales (see below). However, even the poorest waters of the Antarctic Ocean seem to compare favourably in yield with the rest of the world's oceans. Taking Antarctic surface waters as a whole, El-Sayed has calculated that on average 0·89 grams of carbon are assimilated (i.e. extracted from the sea and built into plant tissues) beneath every square metre of surface each day. This is about six times the productivity of the world's oceans as a whole, which another biologist (E. Steeman Neilsen) has estimated at 0·15 $gC/m^2/day$. Taking into account annual distribution and movement of the pack ice and the number of hours of sunshine per year, the productivity figure for Antarctic waters represents an average gross production of $0{\cdot}33 \times 10^{10}$ (3·3 billion) tonnes of carbon per year, about one-fifth of mean gross production in all oceans. So this small ocean, with area about one-twentieth of the world's oceans, accounts for one-fifth of the world's marine production of carbon. The richest Antarctic waters, between Antarctic Peninsula and the South Shetlands, yield 2·76 $gC/m^2/day$, or over eighteen times the mean for the world's oceans.

Phytoplankton reaches its lowest standing crop and productivity values between April and July, when days are short, the sun is low in the sky or absent altogether, sea ice spreads far out from the continent, and plant and animal plankton descends into subsurface layers of the ocean. With the gradual return of the sun in July and August, both standing crop and productivity begin to rise slowly, but remain relatively low until September or October. From October onward both increase dramatically, as phytoplankton returns to the surface and begins to photosynthesize actively and reproduce. The annual "blooming" of the phytoplankton starts in open water north of the pack ice, spreading steadily southward as the ice recedes. Highest values of standing crop and productivity are recorded close to the continent in January, and are followed by a rapid fall through February, March and April as day length and surface temperatures decrease. Even at the height of summer the plants do not exhaust the mineral resources of the richest patches of ocean.

Zooplankton The euphausiids, copepods and other planktonic animals follow a similar pattern of abundance in surface waters. From May to September most animals of the plankton winter

Sponges are often the dominant components of the benthos below the level of severe ice scoup. Here they are associated with other filter feeders, brachiopods, ascidians and polychaetes, together with browsing echinoderms.

A shallow-water assembly of sea-bed animals: nudibranch (centre), spindle-legged pycnogonid or sea spider, sponges, sea anemones and corals.

well below the shallow layer of Antarctic Surface Water, in depths of 250 to 2,000 metres. Thus for most of the winter they are travelling slowly eastward and south with the Warm Deep Current. This is a lean period for surface-feeding birds and seals, which cannot follow the bulk of their prey into the depths, and must hunt as best they can among the thin surface populations. From October onward the zooplankton rises. Like the phytoplankton, its shoals first appear in open water north of the pack ice, and seem to spread rapidly southward as the floes disperse. They may indeed be encouraged by the break-up of the ice, which possibly stimulates diatom production by allowing wind and weather to stir the ocean after its winter calm. Wherever zooplankton is plentiful, the large animals which feed on it begin to accumulate. These include whales, and it has been the business of scientists and whalers over many years to discover where and at what times of the year zooplankton congregates at the surface.

Zooplankton gathers, not surprisingly, where phytoplankton grows thickest, i.e. in inshore and turbulent waters where surface and subsurface layers mix. Some authors suggest that strong winds, due to cyclonic storms and other pressure changes, may help to promote mixing locally, and bring shoals of krill or copepods to the surface. Areas of particular concentration occur along the coasts of Dronning Maud and Enderby Lands between 0° and 60°E., off Queen Mary Land and western Wilkes Land between 95° and 120°E., and off George V Land in 155°E. All these regions lie within the East Wind Drift, the inshore current which carries locally enriched surface waters westward about the continent. There are moderate to high concentrations of zooplankton off Alexander Land and the southern shores of Antarctic Peninsula, but by far the highest concentrations occur in the disturbed, enriched waters off northern and western Antarctic Peninsula and the Scotia Arc. The ice-filled Weddell Sea is a relatively poor area, except for a curiously rich patch of plankton in its southwestern corner. However, where the East Wind Drift hugs the coast southward into the Weddell Sea, it is deflected north by the east side of Antarctic Peninsula and emerges as the strong, northeastward-moving Weddell Sea Current. This carries cold water of unusual richness, with dense phytoplankton and zooplankton, past the South Orkney and South Sandwich Islands. A similar but weaker flow carries enriched water clockwise about the Ross Sea, on the opposite side of the continent.

Practical problems of catching, sampling and handling make it difficult for marine biologists to estimate the standing crop and productivity of zooplankton populations. Results determined independently by British and U.S. expeditions suggest that Antarctic surface waters on average contain the equivalent of eight to ten grams (about one-third of an ounce) of dry animal tissue for each square metre of surface. But this figure must be

Amphipods scattered over boulders and a yellow sponge. The common echinoderm, *Sterechinus*, is also present.

Right: A group of ascidians—sea squirts—spread their syphons to filter organisms from the water.

Limpets (*Patinigera polaris*) and a large starfish browse across a shallow water benthic community of encrusting algae, sponges and hydroids.

BONE

SMITH

BONE

Djerboa furcipes, a common amphipod among shallow water seaweeds. Amphipods are a predominant crustacean group in the Antarctic seas.

Above, right: Emptying the trawl net. Though SCUBA diving helps marine biologists to study shallow water communities, trawling and dredging are still necessary for collecting in deeper water.

A species of Antarcturus from Signy Island, Antarctica. This group of marine isopods are adapted for particle feeding and are common at most depths in the Southern Ocean.

low, for some of the larger animals of the plankton are known to be able to dodge the narrow catching nets which biologists use for accurate quantitative studies. Euphausiids in particular are missing from the reckoning. These large shrimp-like crustaceans grow to seven or eight centimetres long, forming enormous shoals which colour the sea red. J. W. S. Marr, a British biologist who worked for many years on *Euphausia superba*, the common species of the northern Antarctic zone, estimated that this one species alone could reach densities of 29 grams per square metre of surface. Included in the calculations of standing crop, euphausiids must therefore double, treble and even quadruple the estimates for extensive regions of the Antarctic Ocean.

Yield is equally difficult to measure accurately. If euphausiids and other large organisms of the plankton brought standing crop on average up to 20 grams per square metre (doubling the British and U.S. estimates above, though still perhaps on the conservative side), and if annual yield were approximately equal to standing crop (i.e. if the crop effectively doubled itself in the course of the year), productivity over the ocean as a whole would amount roughly to 360 million tonnes of animal material per year. This is necessarily a rough calculation, but the figure is probably of the right order of magnitude, and underestimates rather than exaggerates the resources of the ocean.

If this yield were spread evenly between Convergence and continent, the large animals which depend for their livelihood on zooplankton would work hard and travel far for a square meal. But zooplankton, like phytoplankton, tends to concen-

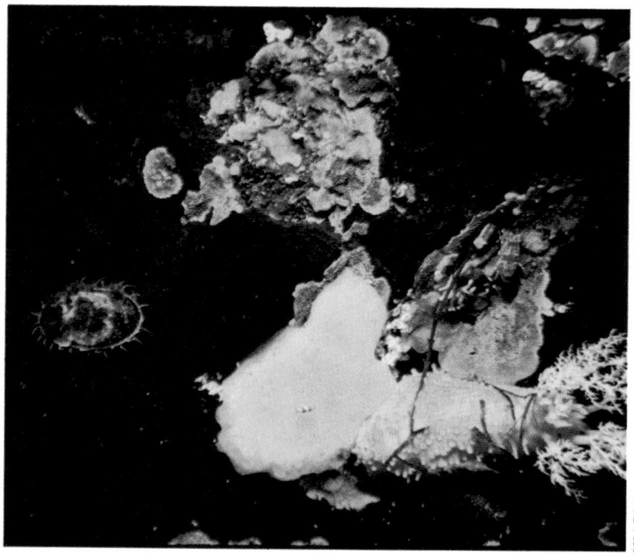

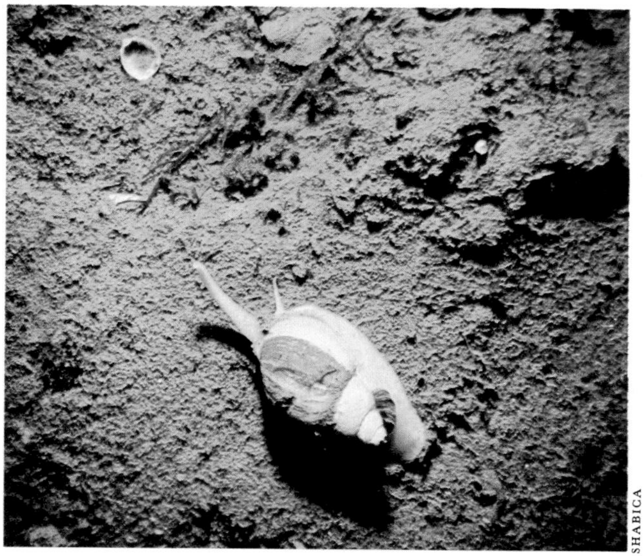

Snail-like gastropod mollusc, collected on a bed of glacial mud close to the shore, in a depth of 14 metres. Many gastropods are carnivorous, killing and feeding on cockles and other fixed molluscs of the sea floor.

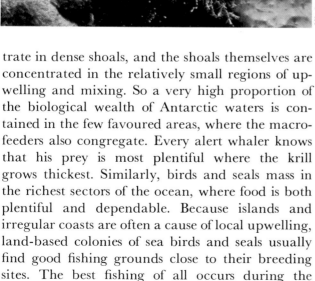

Above, left: Amphipods and limpets (*Patiniger polaris*) wander in an animal forest of cylindrical holothurians, ascidians and encrusting bryozoans. A cliff community in shallow water, 8 metres below tide level, Arthur Harbour, Anvers Island.

Glyptonotus antarcticus, a large marine isopod of the family Idotheidae. With their clawed feet and hinged bodies, isopods crawl with agility in search of their food—mainly organic debris—on the sea floor. Large males are 5 centimetres or more across, females seldom more than 4 centimetres.

trate in dense shoals, and the shoals themselves are concentrated in the relatively small regions of upwelling and mixing. So a very high proportion of the biological wealth of Antarctic waters is contained in the few favoured areas, where the macrofeeders also congregate. Every alert whaler knows that his prey is most plentiful where the krill grows thickest. Similarly, birds and seals mass in the richest sectors of the ocean, where food is both plentiful and dependable. Because islands and irregular coasts are often a cause of local upwelling, land-based colonies of sea birds and seals usually find good fishing grounds close to their breeding sites. The best fishing of all occurs during the annual flush of zooplankton in mid-to-late summer. Both birds and seals have adapted their breeding cycles to fit. The young of nearly all species are growing most rapidly, and need food with the greatest urgency, in the late months of the Antarctic summer—January, February and March.

The plankton feeders

Zooplankton is eaten with relish by a host of molluscs, fishes, birds, seals and whales. Some of these are exclusively plankton feeders. Others vary their diet, feeding on whatever is most plentiful from time to time, and migrating when the plankton descends in late autumn. The diagram on page 43 shows the more important feeding relationships known to exist in far southern surface waters. This diagram is certainly an oversimplification of a complex and intriguing food web, a food web which biologists have been working on for many

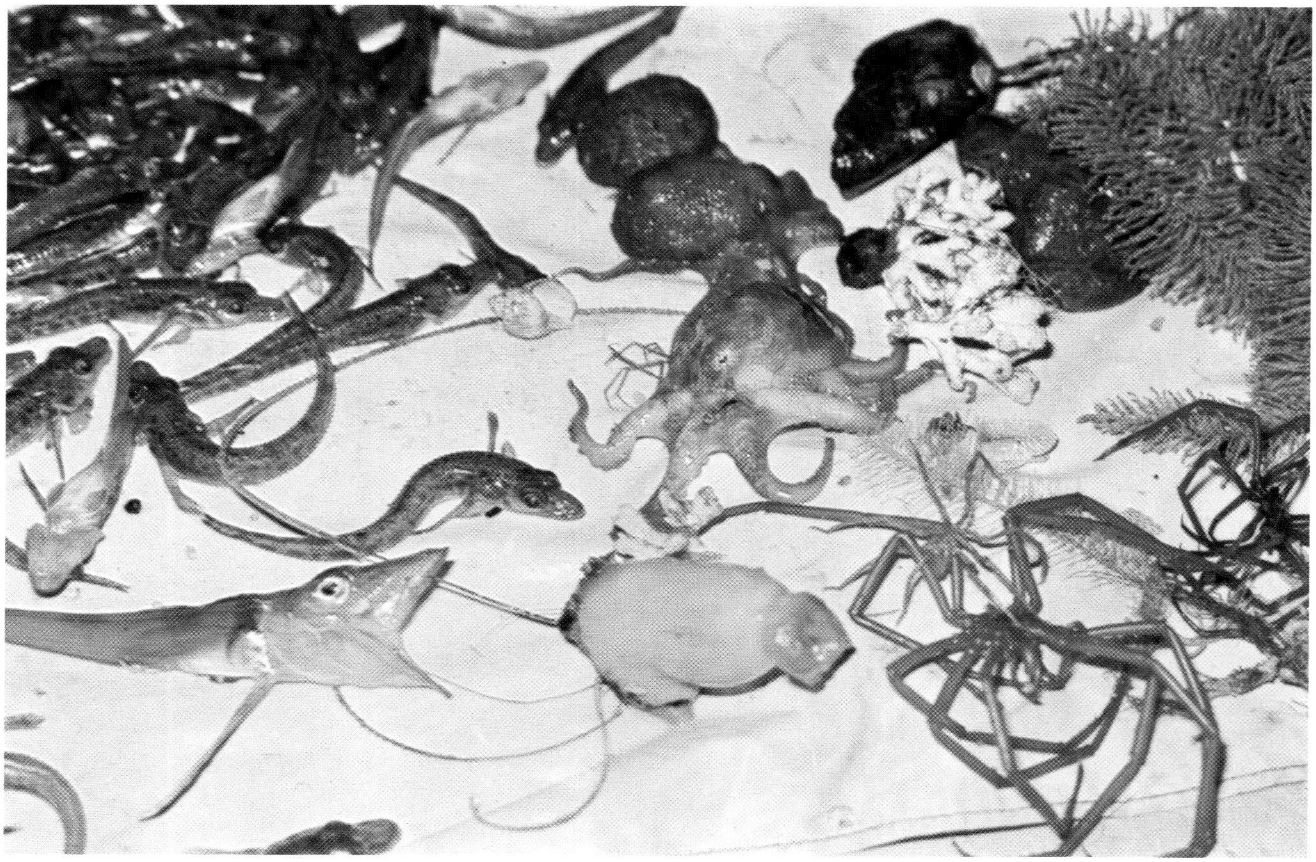

LAWS

This trawl haul includes, in the foreground, an ice fish, *Chaenocephalus aceratus* (left), (centre) a sea anemone, *Zoantharia*, with a sea whip snaking round behind it, and (right) sea spiders, *Pycnogonida*. Behind these are (left) a heap of crocodile fish, *Priondraco evansi*, two octopuses (centre), a hydroid colony, cup coral and (far right) sea fans. Between the octopuses and the sea fans lies a white ectoproct colony and between this and the sea fans is a sea urchin. Another sea urchin is at the back of the picture, behind the second octopus.

years and which they are still trying to unravel.

The most prominent plankton feeders are birds. Thirty-one species of flying sea birds—mostly petrels—and seven species of penguins breed south of the Antarctic Convergence. Practically all feed on the larger animals of the zooplankton, overlapping considerably in their choice of food but specializing to some degree as well (page 88). Many species of birds feed also on fishes and small squid, mostly caught in surface waters. Of the six species of Antarctic seals, one—the crabeater—feeds almost exclusively on planktonic arthropods. It lives mainly among pack ice, and its teeth bear intriguing cusps which help it to filter zooplankton from the water. Among the remaining species, leopard seals feed largely on birds and fish; the rest tend to follow squid and fish into deeper water and ignore the resources of the surface. The six species of Antarctic baleen whales feed almost exclusively on plankton, straining it from the sea with their remarkable filters of whalebone or baleen. Large blue and fin whales take at least one tonne of zooplankton per day. Dolphins and small toothed whales also feed mainly at the surface, but on fishes, squid and seals rather than plankton. Sperm whales, largest of the toothed whales, seem to feed entirely on large squid which they catch in deep water.

Very little is known of the fishes and squid which occupy so important a position in the food web. About ninety species of bottom-living and sixty of surface-living fish have been identified in Antarctic waters. Of those which live on the bottom, about three-quarters belong to the sub-order Notothenioidei, a group restricted to cold waters of the southern hemisphere. Four of the five families within this group—the Nototheniidae (Antarctic cod—unrelated to northern cod and only superficially similar), Chaenichthyidae (ice fishes), Bathydraconidae (dragon fishes) and Harpagiferidae (robber fishes)—are almost entirely limited to Antarctic waters. Other Antarctic fishes include eel-pouts (Zoarcidae), sea snails (Liparidae) and rat-tailed fishes (Macrouridae), hagfishes (Myxini-

Sea spider (life size): a perfect specimen of the rare twelve-legged Pycnogonid *Dodecalopoda mawsoni*, collected in shallow water off the South Shetland Islands.

dae), skates, and a number of bathypelagic (i.e. middle deep-water) fishes.

Fishes of the family Nototheniidae are mostly heavy, sleepy, bull-headed creatures with enormous mouths, which browse among the rocks and mud of the sea bed. The Chaenichthyidae are remarkable in lacking the red blood pigment haemoglobin; some are so colourless as to be almost transparent, as their name "ice fish" suggests. Together with some of the Bathydraconidae, these families are represented at the sea surface by species which spend at least part of the year feeding on plankton. The genus *Trematomus* (Nototheniidae) seems especially to be adapted to surface life. Fingerlings and juveniles of several species attach themselves to the submerged walls of icebergs, and dodge in and out of the crystalline cavities below ice floes, where they feed on diatoms and small crustaceans. Several other nototheniids are now known to form dense shoals in surface waters, and there is increasing evidence that Subantarctic forms (e.g. the southern blue whiting, previously recorded on the continental shelf of Patagonia) also move south to feed on the summer bonanza of zooplankton. Some of these species may be plentiful enough to make commercial fishing a possibility.

Living in very cold water does not seem to place Antarctic fishes at a disadvantage. The enzyme systems responsible for their metabolism and well-being work well at low temperatures, and the fishes are as active, and grow as rapidly, as many species in temperate waters. Different species show different degrees of tolerance to low temperature and to temperature change. Several which are especially adapted to very low water temperatures (e.g. $-1{\cdot}8^\circ$ to $-1{\cdot}9^\circ$C., off the coldest stretches of the continental coast) cannot tolerate change at all; a rise even of 2°C. incapacitates or kills them. This is a hazard which they would not normally meet in the very stable cold of the far south. It could occur in summer on the west coast of Antarctic Peninsula. This may be a reason why the coldest living fishes are restricted to high latitudes. Conversely, several species of the Peninsula which tolerate an annual temperature cycle of 2° to 3°C., are not found elsewhere along the continental coast, though they have spread to the islands of the Scotia Arc where conditions are similar to those of

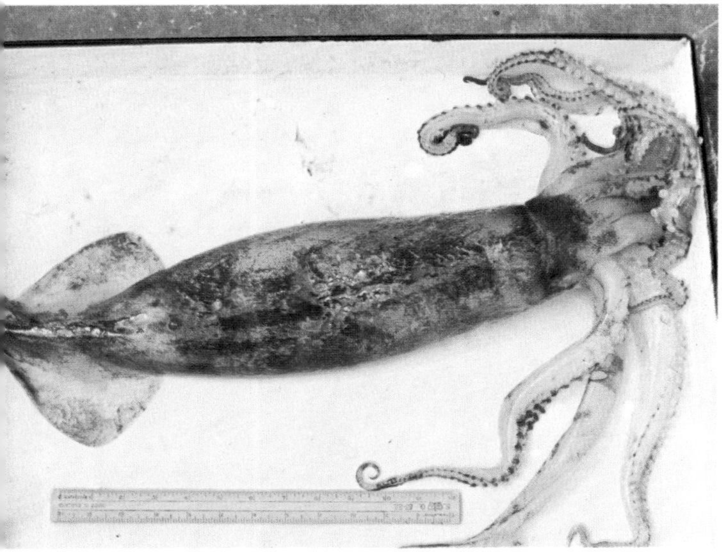

SMITH

Squid or cuttle-fish, a fast-moving relative of the octopus. Though squid are plentiful throughout Antarctic seas, they are rarely caught by man, and very little is known of their biology. They are a favourite food of seals and large sea birds.

BONE

the Peninsula. The British marine biologist N. B. Marshall has estimated that, of all the Antarctic fishes found close to the continent, only about one-third are completely circumpolar. About one-sixth are confined to the west side of the Peninsula and Scotia Arc islands, and the remaining half to the colder mainland coasts of the continent.

How do polar fishes avoid freezing in their ice-cold environment? As in all other bony fishes, the body fluids of polar species are less salty than the surrounding sea. Theoretically they should freeze at slightly higher temperatures than sea water, and it has been an interesting problem to discover how polar fishes manage to remain unfrozen at environmental temperatures below $-1°C$. Species which live well below the sea ice apparently have little problem. Though their blood would normally freeze at environmental temperatures, they are capable of supercooling their tissues slightly—by about $0{\cdot}7°C$.—and remain safe so long as they do not come into direct contact with ice. If, by chance, ice crystals touch their skin or gills, body fluids begin to freeze and the fishes die immediately. In

EVERSON

Harpagifer bispinis, a coastal robber-fish of Antarctic waters, which is also found north of the Antarctic Convergence. The spines can be erected when the fish is threatened, and their bony points are an effective deterrent against predators.

Antarctic cormorants feeding together among a shoal of plankton or small fish, South Orkney Islands. Cormorants usually hunt alone, diving to depths of 10 metres or more for their food.

BONE

Trematomus newnesi, often found in shoals just below the sea ice in winter, are able to survive in contact with ice at temperatures of −1·8°C. for one hour or more. Many other Antarctic fish survive only in ice-free water; direct contact with ice crystals causes their tissues to freeze.

KOOYMAN

Trematomus bernacchii, living among feathery strands of a fan coral and a colonial hydroid (probably *Tubularis*), on the sea bed in McMurdo Sound. This species is common down to 300 metres in coastal waters.

winter their fluids contain a slightly higher salt concentration, to help in coping with the slightly lower temperatures. Fishes which live among ice—immediately under floes or in the sub-surface crystalline cavities of fast ice—have the special protection of a protein-carbohydrate "antifreeze" compound in their blood. This curious substance prevents ice crystals from forming in their tissues at the very lowest temperatures reached by sea water, and the fishes remain safe in their icy caverns.

Remarkably little is known of the squid inhabiting southern polar waters. Though practically every whale, seal, penguin and large petrel feeds on them from time to time, man has found them strangely elusive. Their large eyes see well in the dim submarine light, and they move fast enough to avoid advancing nets. Squid are occasionally hauled in almost by accident in nets and on water sampling bottles and thermometer rigs. They seem to occur at all depths; some may migrate annually and daily with the plankton, while others remain close to the sea bed in shallow or deeper water. Though clearly abundant and a major predator of other creatures, their biomass, productivity, breeding habits, and way of life are unknown.

Chaenocephalus aceratus, an ice-fish of the family Chaenichthyidae. This fish, like other members of its family, lacks haemoglobin. Its gills are white, and its blood colourless. Adequate supplies of oxygen are carried in the blood plasma.

BONE

KOOYMAN

Nototheniid fish resting among yellow-white clumps of coral (*Alcyonium*) and athecate colonial hydroids. Fish up to 50 centimetres long are abundant in shallow coastal waters about the Antarctic continent and Peninsula.

Notothenia gibberifrons, a species widespread in coastal waters around the continent and throughout the Scotia Arc. Like most other small, bottom-living fishes, this species browses on the live animals of the sea bed.

BONE

MANNERING

The Antarctic Islands

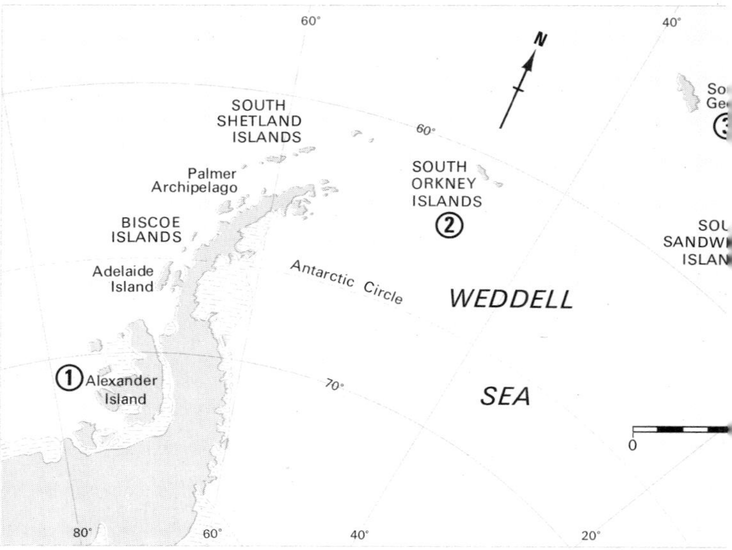

1 Continental islands (e.g. Alexander Island) have only recently emerged from ice cover. Their soils are poor and immature, windswept, and dry during the short summer growing season. Only lichens and scattered moss clumps survive; 99 per cent of the snow-free areas are bare of vegetation.

Vegetation on continental, maritime and periantarctic islands. The three photographs show vegetation typical of the areas keyed on the map.

Previous page: Beaufort Island, a typical continental island. Moraine remnants high on the shoulders of the island indicate that a continuous sheet of ice shelf once linked Beaufort Island to mainland Antarctica.

Mean monthly temperatures at selected Antarctic stations. Continental stations are coldest, both in winter and summer, and have the greatest annual range. Periantarctic islands have the mildest and most uniform climates. Coastal stations and maritime islands are intermediate.

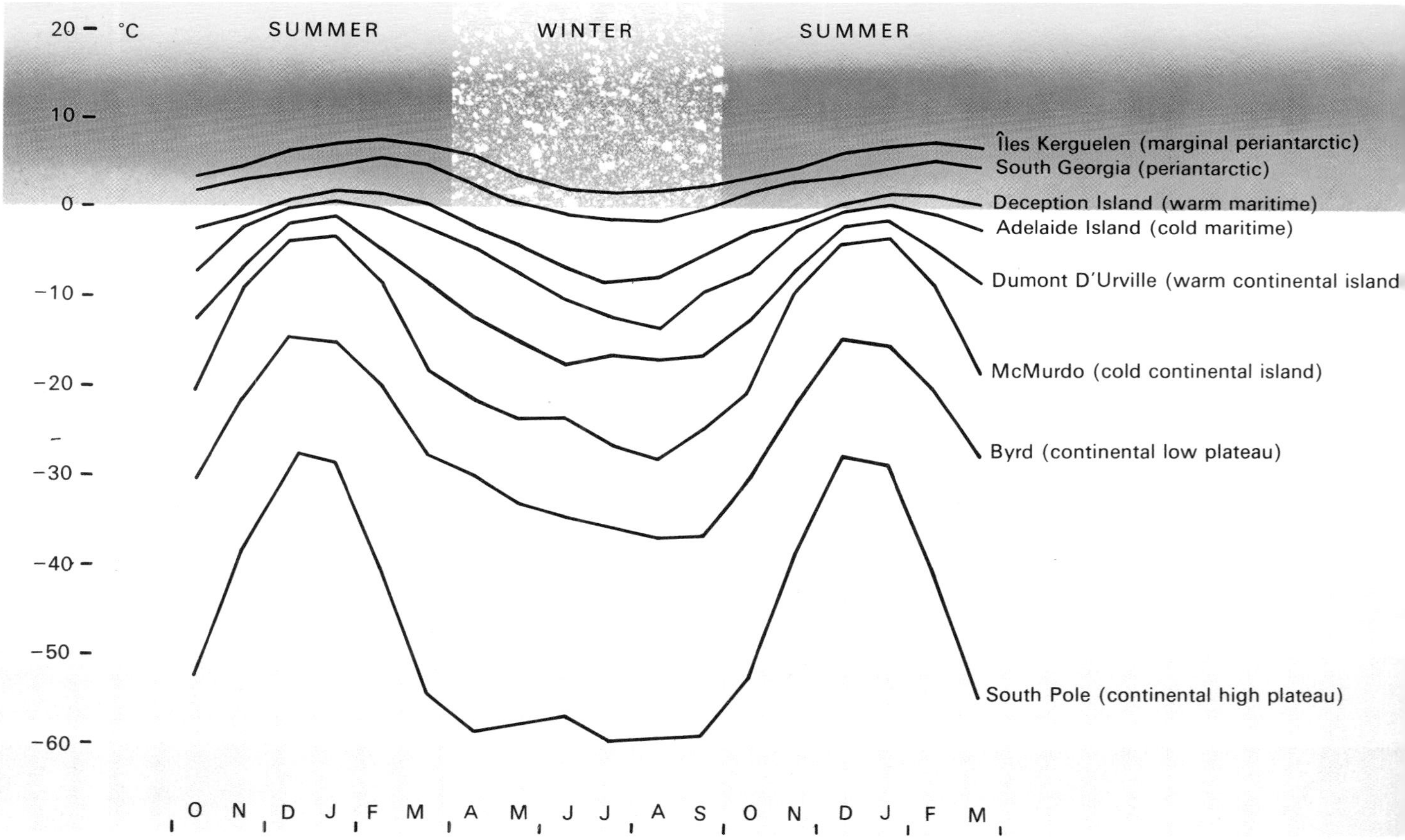

2 Maritime islands (e.g. South Orkneys) have patches of older, more mature soils with higher organic content, a damper climate, and longer growing season. Mosses, lichens, and two species of flowering plant form continuous though patchy carpets of vegetation.
3 Above, right: Periantarctic islands (e.g. South Georgia) are densely covered with vegetation to the snow line.

Continental islands. Ross, Beaufort and Franklin Islands are volcanic islands in the Ross Sea. Mount Erebus, the main peak of Ross Island, remains an active volcano.

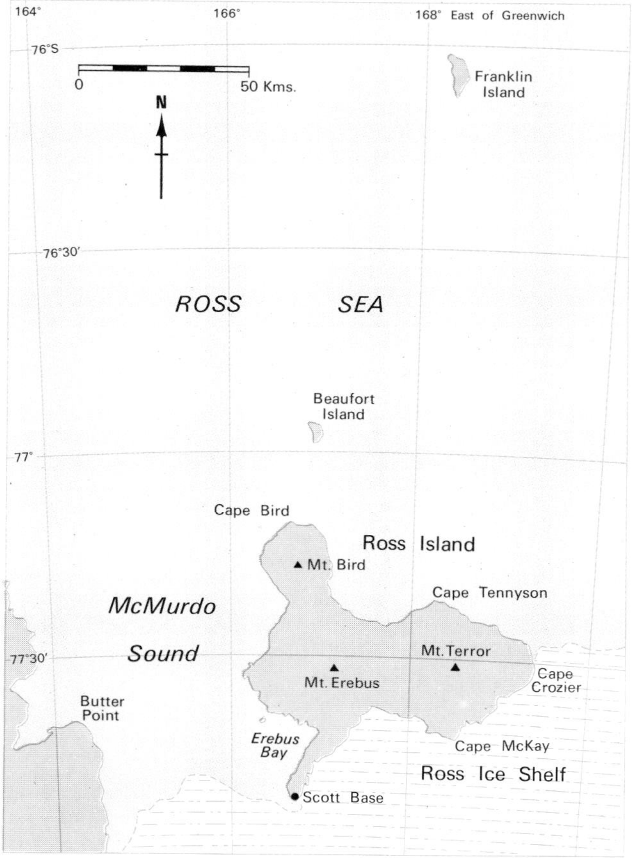

The islands of Antarctic waters fall into three ecological groups—continental, maritime and periantarctic. *Continental islands* lie close under the lee of mainland Antarctica, and are dominated climatically by the continent. Ecologically they are part of the continental coast, with poor soils and sparse desert vegetation. *Maritime islands* stand far enough from the continent to be surrounded by open water in summer, though linked by pack or fast ice to their cold neighbour in winter. They have a slightly longer growing season than the continental islands, usually more summer rain or snow, and are correspondingly greener. *Periantarctic islands* stand free of sea ice throughout the year, except for small amounts which may form locally in sheltered harbours overnight. Though most periantarctic islands are ice-capped, at sea level their summers are long and warm, their winters relatively mild. A long growing season encourages lush tussock grasses and a variety of herbs, which clothe the lower slopes of the islands in green vegetation.

Continental islands

For all practical purposes continental islands are part of the mainland close by. Those lying furthest south have the harshest climate, chilled both in summer and winter by cold air from the interior of the continent, and cut off from the warmth of the sea by persistent ice. Coldest of all are the far southern islands linked by shelf ice to the mainland coast. Alexander Island, at the base of Antarctic Peninsula, is virtually ringed by broad piedmont ice shelves. Thurston Island, only recently recognized as an island, is connected to the Eights Coast of Ellsworth Land by an ice shelf 50 kilometres wide. Standing in a region of heavy snowfall, its high mountains are almost overwhelmed by huge fields of snow and ice. Ice-ringed continental islands have little to offer wild life. Without beaches, they are inaccessible to penguins or seals. Little or nothing grows close to sea level, and only the peaks are likely to be colonized by snow petrels, Antarctic petrels, and other crevice-nesting sea birds.

Beaufort and Franklin Islands, once active volcanos, but now silent and heavily ice-capped, stand isolated in the southwestern Ross Sea. For all but a few days each year they are encircled by pack ice two metres or more thick. Beaufort Island (page 61) was once yoked to the mainland by an extension of the Ross Ice Shelf, which may well have surrounded Franklin Island, too. The climatic changes which broke the link also removed many metres of ice from the thickness of the cap on top of the island. With some of the weight lifted from its shoulders, Beaufort Island has risen several metres from the sea. Recently emerged beaches provide well-drained nesting grounds for many thousands of Adélie penguins and hundreds of McCormick's skuas. Sea ice close to the islands is stable enough to support large emperor penguin colonies, which form in winter and disperse with the sea ice in summer (page 42).

Ross Island, a larger volcanic mass flanking McMurdo Sound in the southwestern Ross Sea, stands in a dry corner of Antarctica. Its ice cap, once thick and continuous, is ablating rapidly. Wide expanses of ice-free moraine and beach have emerged recently—perhaps in the past two or three hundred years—to expose hundreds of square kilometres of dry desert pavement and rubble. Mount Erebus, the main peak of the island, is an active volcanic cone 3,794 metres high. First seen in January 1841, by James Clark Ross of the Royal Navy (page 159), it treated its discoverers to a memorable firework display. Later explorers who based their expeditions in McMurdo Sound found Mount Erebus a quiet neighbour—even a helpful one. For meteorologists of Scott's and Shackleton's expeditions, its smoke plume indicated the strength and direction of upper winds, and helped in forecasting the weather. McMurdo Sound, one of the two most southerly stretches of water in the world, freezes to a depth of three to four metres each winter. Because of strong winds, currents, and pressure from the slowly-advancing Ross Ice Shelf, the sea ice is locally unstable and often disperses in patches as early as October. Polynyas (page 40) forming in the Sound and off Cape Crozier make it possible for penguins and seals to breed in these far southern latitudes. The world's southernmost Adélie and emperor penguins, McCormick's skuas and Weddell seals breed along the shores and bare coastal slopes of Ross Island.

Milder and richer continental islands lie off the coasts of Adélie and Wilkes Lands, in latitudes 65° and 66°S. These, too, have emerged recently from under the retreating edge of the ice cap, which still rests lingeringly on some of them, or stands as tall, slowly calving ice cliffs a few hundred metres to the south. On one such group of islands the French explorer Dumont d'Urville raised the tricolor of France in January 1840, to the astonishment of a group of small black and white penguins standing by. The coast he named Terre Adélie, in honour of his wife. The penguins, some of which accompanied him as specimens back to France, took their name from the coast. Adélie penguins are now familiar as the common circumpolar species—Antarctica's best known and most popular animal.

Winters are cold on these continental fringe islands; the sea is partly or completely ice-covered, and cold air rolls down the long slopes from the plateau above. Summers are relatively mild. From November onward there is a rapid warming as the sea ice disperses and the sun climbs higher. Mean December and January temperatures lie very close to freezing point, and November and February means are only slightly lower. So there is a growing season of two to four months. Thick beds of peaty moss build up in sheltered corners of the islands, and a dozen species of birds nest on them. Snow, Antarctic, pintado and Wilson's petrels nest in crevices and sheltered places among the rocks, with large colonies of giant petrels, Adélie penguins, McCormick's skuas, and possibly emperor penguins and Weddell seals on the beaches and inshore ice.

Maritime islands

These lie in a zone slightly to the north of the warmest continental islands, most of them concentrated in the sector close to Antarctic Peninsula where—with the Peninsula itself—they form a climatic and ecological "maritime province". Jutting far from the continent, Antarctic Peninsula and its fringing islands lie in the track of eastbound cyclones or depressions, which originate in the southern Pacific Ocean and swirl obliquely through Drake Passage toward the continental coast. Their warm air carries moisture, which condenses on cooling to form the constant pall of grey cloud characteristic of the province. In winter much of the moisture falls as heavy snow, in summer as drizzle, sleet or wet snow, and many of the maritime islands are thickly invested with caps of compacted snow and ice (page 64). Maritime islands elsewhere about Antarctica have a similar climate and capping of snow.

Fast ice surrounds all the maritime islands from May or June to October, persisting as pack ice for a further four or five months of the year. But winters are seldom so cold as those of the continental islands; mean winter temperatures usually remain above −10°C. Summers are positively warm in comparison with islands further south; mean temperatures remain above freezing point for two, three or four months of the year. From October or early November to April, plants and animals of land and fresh water live and grow actively. Organic soils develop and hold moisture through the summer. Thick beds of moss and lichen provide a springy turf, with Antarctica's two species of flowering plants flourishing among them. Freshwater lakes have time to thaw out completely and warm through, often becoming green with rich algal growth. Sea salts blown by the wind, and droppings from thousands of burrowing

Cape Armitage, Ross Island. Dominated by the volcanic peak of Mount Erebus, Ross Island is the southernmost point in Antarctica which can be reached by ship. McMurdo base (US) lies to the left of the cape, Scott base (New Zealand) on the far right, near the junction of island and ice shelf.

LEWIS SMITH

Lichens *Xantharia candelaria* and *Parmelia sp.* encrusting the rocks of Adelaide Island. Lichens and scattered moss clumps are the characteristic vegetation of continental islands and colder maritime islands.

Right: Antarctic Peninsula, with off-lying Adelaide Island, the many islands of the Palmer Archipelago and the South Shetland islands, together form a series of curved, almost parallel mountain chains. All are heavily ice-capped, and the Peninsula is deeply dissected by glaciers on both flanks. The sedimentary soils of Alexander Island, which lies in the crook of Antarctic Peninsula, have yielded a wealth of fossil material. The north-western flank of the Peninsula extends into the Maritime climatic zone. The eastern side, chilled by the cold circulation of the Weddell Sea, is heavily icebound and has a continental climate.

Many of the South Shetland islands bear a load of ice which is constantly renewed by heavy winter snowfall. Active glaciers sweep down between the mountains to sea level, leaving very little flat ground exposed.

BURLEY

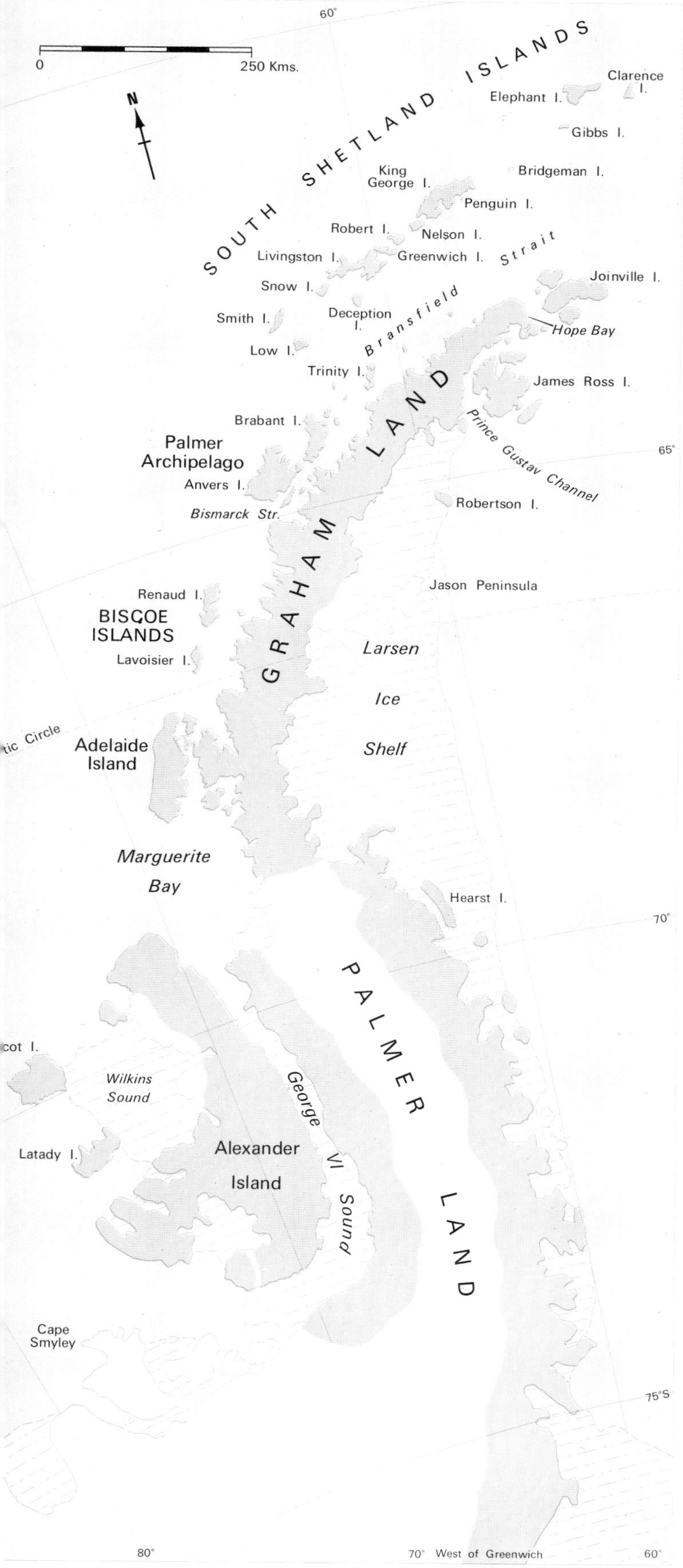

Steam rises from a sun-warmed moss carpet, Elephant Island, South Shetlands.

"Moss peat" over 1 metre deep, Elephant Island, South Shetlands. Similar accumulations, probably representing several hundred years growth, are recorded on the South Orkney Islands, and in sheltered valleys along the Peninsula coast.

YOUNG

Chinstrap penguins nesting on fresh basalt flows on the outer coast of Deception Island. Volcanic dust from recent eruptions covers the snowbanks.

Top, right: Clouds of steam rising over Deception Island during an eruption.

Right: *Colabanthus quitensis*, a small flowering plant of the maritime Antarctic, forming thick cushions on a moist slope of Deception Island, South Shetlands.

sea birds, contribute minerals to the soil and encourage growth. Many areas of the maritime islands appear green where permanent ice and snow are absent.

Coldest of the maritime islands are Adelaide Island, the Palmer Archipelago, and other islands flanking the west coast of Antarctic Peninsula. Still strongly under the influence of the continent, especially in winter, they are only slightly milder than continental islands immediately to the south. Substantial ice caps and piedmont fringes help to cool them. The South Shetland Islands, an extensive archipelago running parallel with the west side of the Peninsula, are slightly milder. Geologically the South Shetlands form an arc of submerged mountains, of which only the snow-capped peaks and rocky shoulders tower above the sea. Three large islands—Livingston, King George and Elephant—dominate the group; eleven smaller islands and many snow-capped rocks complete the chain.

Three of the islands, Penguin, Bridgeman and Deception, are volcanic and relatively new. Bridgeman Island was reported active in 1821 and again in 1839, but is now quiet. Deception Island, a flooded volcanic crater known as a safe harbour to generations of sealers and whalers, rumbled into activity in 1967, to the disquiet of scientists manning its research stations. Dust and ashes rose high into the atmosphere, and a new island nearly 1,000 metres long appeared in the crater harbour. Two years later Deception exploded again; a further eruption cut fissures in the flank of its highest mountain, on the east rim of the island, and hot gases and magma melted part of the remaining ice cap. There was a further, smaller eruption in August 1970. Fumaroles close to the sites of the eruptions had for long warmed the ground, encouraging the growth of green moss patches among the permanent snows. Normally the South Shetlands are peaceful, providing bases for millions of

BAS

LEWIS SMITH

BONE

Moss-covered slopes and orange lichens provide a backdrop for brown skua in the maritime zone, South Orkney Islands.

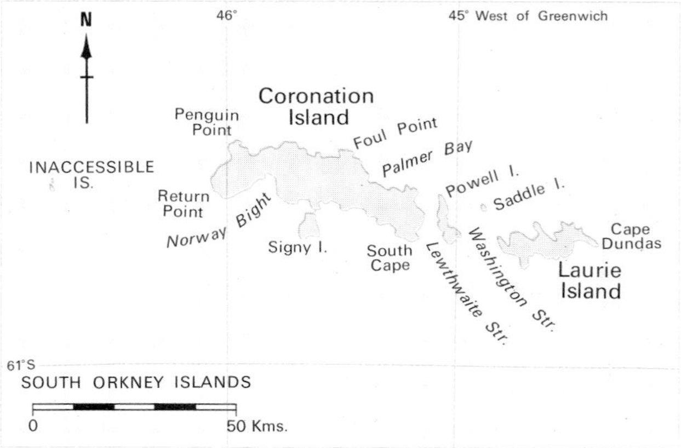

The South Orkney Islands lie north and east of Antarctic Peninsula. A series of depressions passes over them, giving them a damp, cloudy climate with heavy snow in winter and sleet in summer. Cold air and sea currents flow past continuously, and pack ice surrounds them for ten months of the year.

LEWIS SMITH

Movement of particles within wet soil, caused by alternate freezing and thawing, sorts fine and coarse material into a pattern of stripes on sloping ground. Constant surface disturbance discourages plant life.

On flat ground, similar movements in wet soil cause the formation of rings or polygons. Plants cannot establish on the active centres of the rings.

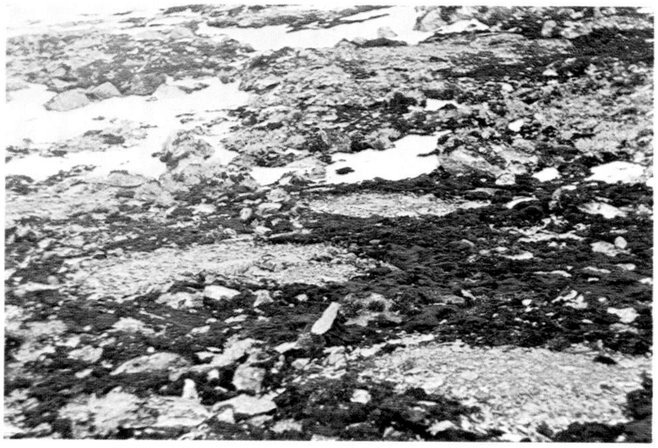

EDWARDS

penguins, petrels and seals. Soils are seldom developed beyond a primitive organic phase, but thick beds of moss-peat have been reported from many of the islands, and lichens, moss and algae flourish in the damp summer atmosphere.

The South Orkney Islands, a smaller and more compact group, lie 700 kilometres east and slightly north of the Peninsula tip. They are no warmer than the South Shetlands, for the cold Weddell Sea Current (page 24) laps round them, bringing fast ice or pack ice from March to January each year. Coronation and Laurie Islands, largest of the group, are heavily burdened with permanent ice caps. Signy Island, a smaller, low-lying triangular island in their lee, has lost all but a remnant of the ice cap which once covered it. A saddle of bare ground on Laurie Island houses Orcadas, the oldest-established human settlement in the Antarctic region. Founded by the Scottish National Expedition in 1903, it has been manned continuously by meteorologists of the Argentinian government, and now presents a unique climatic record spanning seventy years. Adélie and chinstrap penguins form enormous colonies. Over one million chinstraps are said to breed annually on Laurie and neighbouring Saddle Island, and about five million Adélies nest on Laurie Island alone. The rolling coastal plains of Signy Island, formed of ancient glacier-smoothed schists and amphibolites, are crossed by bands of white marble. Fierce post-glacial weathering has produced a variety of soils, which in summer are practically snow-free but remain saturated with ground water. Signy Island's soils, vegetation, terrestrial and freshwater animals, seals and petrel

LEWIS SMITH

Signy Island, standing in the rain shadow of other South Orkney islands, is relatively ice-free, with wasting glaciers and exposed moraines. Varied metamorphic rocks, including marble (white knoll on left) provide varied soils and patterns of vegetation.

Thousands of Chinstrap penguins nest on Frederiksen Island, South Orkneys, an island in the maritime Antarctic zone. Steep sloping cliffs and recently exposed moraines have a thin, incomplete covering of mosses and lichens.

Though deep, stable soils form in well drained corners of the maritime islands, they lack humus and seldom mature beyond primitive "brown soil" stage. Signy Island, South Orkneys.

Polytrichum, a moss and *Usnea*, a branching lichen, forming a thin carpet over volcanic ash on Bellingshausen Island, South Sandwich group. Hot volcanic gases arise from the basalt ridge behind.

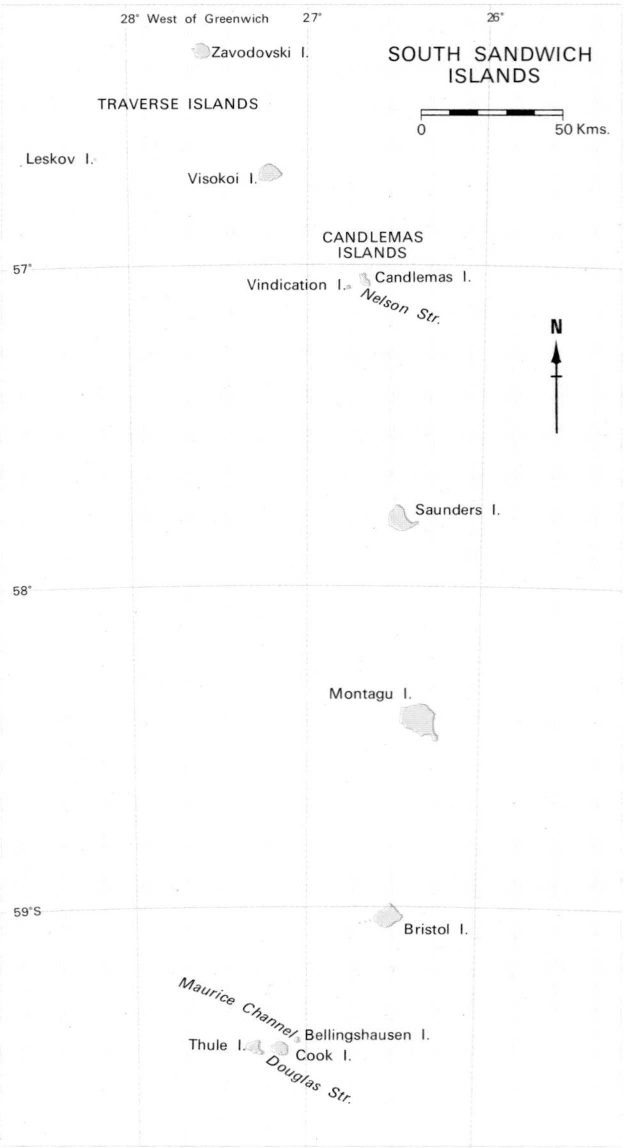

The South Sandwich Islands, a north-south chain of islands at the eastern end of the Scotia Arc. Many of the islands are active volcanoes: most are heavily ice-capped, and surrounded by sea ice for eight to ten months each year.

colonies have been studied for many seasons by British scientists from a permanent base on the island (page 24).

The South Sandwich group is a chain of eleven volcanic islands, forming a crescentic eastern end to the Scotia Arc 350 kilometres long. The southernmost islands, Thule, Bellingshausen and Cook, are 1,000 kilometres east of the South Orkneys and a little further north. The largest, Montagu Island, is roughly 12 kilometres in diameter, and rises to an ice-capped peak 1,370 metres high. The remaining islands are mostly single or complex cones, heavily glaciated, with vertical rock or ice cliffs at sea level. The southern islands generally carry thicker ice than the northern ones. Most of the islands are still volcanically active. Sealers, whalers and other visitors have from time to time reported columns of steam and sulphurous fumes rising from them, and submarine disturbances close by. In 1908 the captain of a visiting whaling ship was poisoned by fumes from the rocks of Zavodovski Island. Only Cook, Montagu and Vindication Islands seem to be entirely inactive and cold.

In almost every way the South Sandwich Islands are less hospitable than the South Orkney

Chinstrap penguins nesting at the edge of a fumarolic lake, Cauldron Pool, on Candlemas Island in the South Sandwich group.

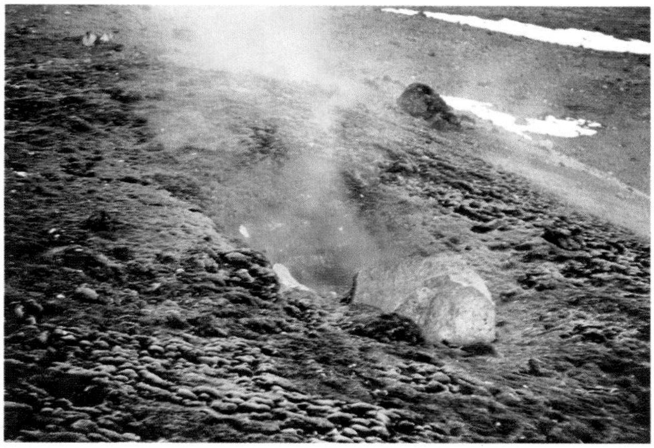

Warmth and moisture near fumaroles on an island in the South Shetlands group encourage a rich growth of mosses. Elsewhere the thin volcanic soils of the group support only meagre vegetation.

and South Shetland groups. There are fewer beaches, or points of easy landing for man or beast, and only a very small proportion of land in the group as a whole is free of ice. However, some of the bare coastal plains, notably those on Zavodovski and Bellingshausen Islands, house enormous colonies of penguins. Zavodovski Island alone is reported to support as many as 14 million breeding chinstraps, and other species of penguins—Adélie, gentoo, macaroni and king—have been seen in large numbers about the group. Petrels breed in thousands on the steep cliffs; brown skuas, dominican gulls and cormorants are plentiful, and giant petrels breed on some of the islands. Five species of seals are found in waters close to the islands, but only elephant and fur seals are known to breed there.

The ashy soils of the South Sandwich group dry out quickly and, except in damp, steam-heated areas close to the volcanic vents, do not support lush vegetation. The green alga *Prasiola* is widespread, but moss and lichen communities are sparse, and the islands have a barren, forbidding appearance. No permanent bases have been established on them; though whalers often hunted the rich waters close by, there are no good anchorages and few scientists have had opportunities to work on this extraordinary group.

Other islands, even more remote and unstudied, lie scattered in the maritime zone. Peter I Øy, in the Bellingshausen Sea, is a solitary peak 22 kilometres long and less than half as wide, ice covered and rising to 1,220 metres. Usually surrounded by dense pack ice, it is reported to have very little flat ground or vegetation. It supports only a single small breeding colony of 90 to 100 Adélie penguins, one or two pairs of chinstraps, their attendant McCormick's skuas, and large numbers of cliff-nesting Antarctic fulmars—possibly also snow petrels, Wilson's petrels and other species. The Balleny Islands, off the northern coast of Victoria Land, are a chain of three large and two small islands lying athwart the Antarctic Circle. The largest and southernmost, Sturge Island, is 43 kilometres long and rises to over 1,500 metres. All the islands are ice-covered and surrounded by pack ice for much of the year. There is a single large colony of Adélie penguins, numbering several

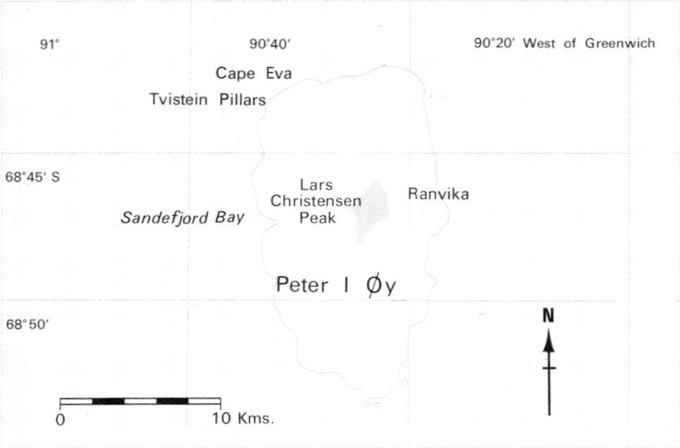

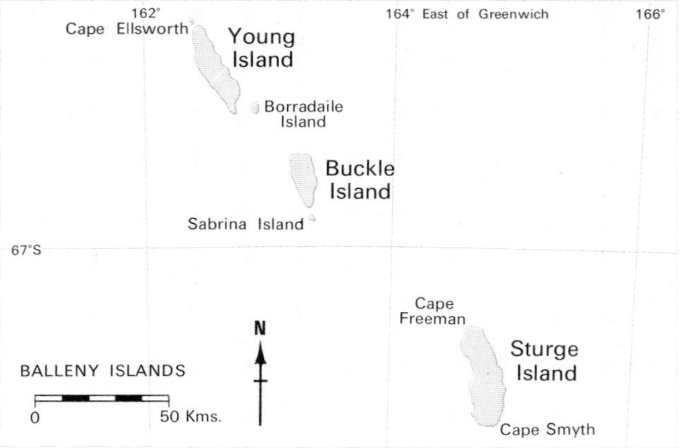

The Balleny Islands, a chain of steep mountainous peaks off the coast of George V land.

Above, left: Peter I Øy, standing alone in the Bellingshausen Sea, has rarely been visited since its discovery in 1821.

LEWIS SMITH

Pintado petrels nest in the cliffs of many maritime islands. Local accumulations of nitrates from their droppings encourage growth of algae and lichens.

Right: Grass-covered slopes and a crater lake in the interior of South Georgia. In the background, ice-covered Mount Sugartop rises to 2,300 metres.

Periantarctic islands north of the pack ice have ice-free shores where intertidal seaweeds flourish and animals can loaf. Bay of Isles, South Georgia.

STONEHOUSE

LEWIS SMITH

thousand pairs, and a number of smaller colonies on tiny patches of bare ground. A few pairs of chinstrap penguins have been reported, apparently trying to nest among the Adélies, and the cliffs of several islands house colonies of snow and pintado petrels, possibly also Antarctic fulmars and Wilson's storm petrels. Mosses, lichens, and surprisingly rich-looking soils occur on Sabrina Island. Small numbers of Weddell seals and a single bull elephant seal have been reported.

Scott Island, an isolated stack 650 kilometres east of the Balleny Islands, remained undiscovered until 1902. Only 400 metres long and 40 metres high, it is largely ice-mantled and usually surrounded with loose pack ice. There does not seem to be enough flat ground near sea level even for penguins to foregather. Flying birds, probably some of the smaller petrels, are reported to rest high on cliff ledges beyond reach of the waves.

Periantarctic islands

These are the warmest Antarctic islands, lying between the northern limit of pack ice and the Antarctic Convergence. At sea level they are damp and green; mild in summer, and cold—but never very cold—in winter. Among their high peaks they are fiercely Antarctic, with violent blizzards and constant sub-zero temperatures. Cold, turbulent winds sweep frequently from the mountains, bringing chill to coastal plains and whipping inshore waters to frenzy. South Georgia, typical of the periantarctic zone, is a whale-shaped island 170 kilometres long and up to 40 kilometres broad, with a blunt eastern snout and a slender western tail. The peaks of its central massif rise in a spectacular ridge to over 2,700 metres. Usually enveloped in cloud, the mountains appear with startling effect during occasional spells of anticyclonic weather, when all is clear and still. Geologically, South Georgia is a fragment of the Scotia Arc, an extension of the Andes far off course in the Southern Ocean.

Warmed by the sea throughout the year, the lowlands of South Georgia are frozen and snow-covered in winter but wet from October to May, when mean temperatures remain above freezing point. Precipitation is heavy, plastering the ground with wet snow and sleet, or soaking it with rain. There are good, deep organic soils with mature layered structure. Meadows of tussock grass line the shore, giving way to rich and varied communities

Braided glacial stream crossing a tussock-clad lowland moraine in one of South Georgia's fiord harbours.

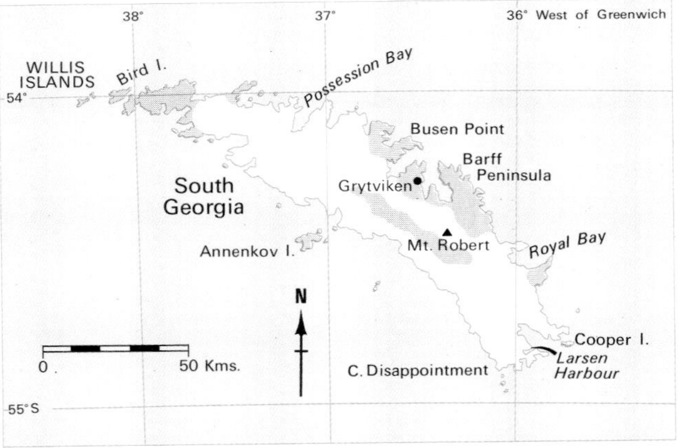

Left: South Georgia is a large island 170 kilometres long, rising to over 2,700 metres. The deeply indented bays and cliffs are the home of many millions of sea birds, and the waters to the east were once the world's richest whaling grounds.

Right: Elephant seals loafing in coastal tussock grass, against a background of moss moorland with shorter grasses. South Georgia.

of flowering plants (twenty species), ferns, mosses, lichens and algae on higher ground. The northern side of the island is warmer than the southern, with more ice-free ground and a higher snowline in summer. South Georgia is the home of two dozen species of breeding birds, many of which nest in immense numbers. Several species nest in, under or on top of the tussock grass, and are unknown on islands south of the pack ice fringe where tussock does not grow (page 24). Its beaches are the rendezvous of many thousands of elephant seals and fur seals, now restored in numbers after man's depredations of the nineteenth century. Smaller numbers of leopard and Weddell seals also breed on the island.

South Georgia has had a permanent though rapidly changing human population since 1904, when the first whaling station was opened at the head of one of its deep natural harbours. So successful was local whaling that other stations were started, and the wreckage of half a dozen rival establishments, in varying stages of decay, now adds to the interest—if not to the beauty—of this fascinating island (page 158). The last of the factories closed in 1965. Whalers had little leisure, but they imported reindeer from Norway, sheep,

The Willis Islands, Bird Island and mainland South Georgia. Albatross colonies cover the steep foreground slopes.

Big Ben, the main ice-covered massif of Heard Island, from the volcanic cliffs of Laurens Peninsula.

horses, upland geese and mallards from the Falkland Islands, and possibly other animals to feed on South Georgia's rich pastures and provide sport and fresh meat. Only the reindeer survive, in two herds on Barff Peninsula and Busen Point Peninsula, which now number several thousand animals. Earlier visitors, probably sealers, introduced rabbits and—unintentionally—brown or Norwegian rats. Rabbits survive only on one offshore island. Rats are now widespread in lowland South Georgia. Their effects on sea bird stocks has never properly been investigated, but they are likely to have decimated lowland breeding petrels and possibly to have driven a number of species—mollymawks, pipits and others—to breed only on offshore islands.

Heard Island is a tiny, tortoise-shaped island in the Indian Ocean, with a peninsula head, high-domed back, and long sand-spit tail. Only 43 kilometres long and half as wide, it rises startlingly to 2,765 metres. Big Ben, its central cone, is an ice-capped volcano with active fumaroles and a spectacular crater. Heard Island arose with its neighbour Macdonald Island from an extensive submarine platform, which it shares with Îles Kerguelen. Marginally colder than South Georgia throughout the year, Heard Island stands in prevailing westerly winds and sea currents, both of which have a cold southerly component. Solidly plastered with thick snow and ice, it has only a few patches of open ground available for plants and animals to colonize. Tussock grass *(Poa cookii*, similar to but specifically different from South Georgia's *Poa flabellata)* fringes the beaches, but coastal vegetation generally lacks the variety and lushness of the warmer Subantarctic islands. Four species of penguins—kings, gentoos, rockhoppers and macaronis—nest on the beaches and exposed scree slopes, and some fourteen species of petrels, skuas, terns, cormorants and other flying birds divide the remaining ground between them.

Macdonald Island is a smaller island 38 kilometres west of Heard Island, only recently landed on for the first time and described in some detail by Dr. G. M. Budd. With outlying Flat Island, Meyer Rock and Needle Island, Macdonald probably forms part of a volcanic cone now breached and partly destroyed by the sea. It is made up of lavas and tuffs, part-covered by luxuriant growth of tussock grass, cushions of *Azorella selago* and Kerguelen cabbage, *Pringlea antiscorbutica,* a large-leaved flowering plant. The northern half of the island, with a total area of 1·3 square kilometres, is a sloping plateau rising to about 120 metres and

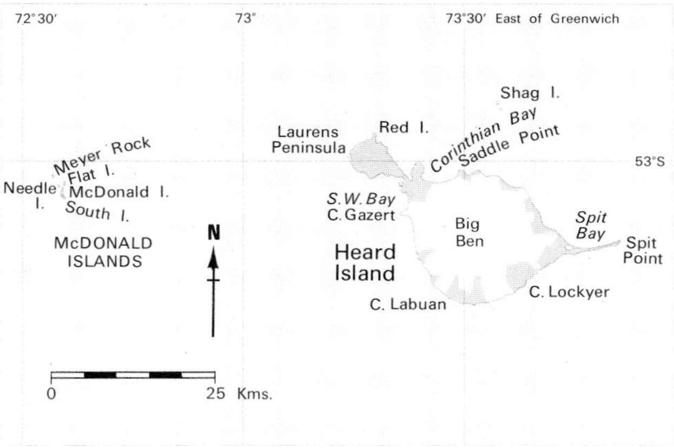

Heard and Macdonald Islands stand, together with Îles Kerguelen, on a vast submarine platform in the southern Indian Ocean. Heard Island towers to over 2,700 metres and is volcanically active. Macdonald Island is a small, cold remnant of volcanic origin 38 kilometres to the west.

YOUNG

The deeply dissected coastlines of Îles Kerguelen provided fine natural harbours for sealers and whalers during the nineteenth century. Rabbits and other browsing animals, introduced to provide fresh meat, stripped vegetation and soil from many of the islands and peninsulas.

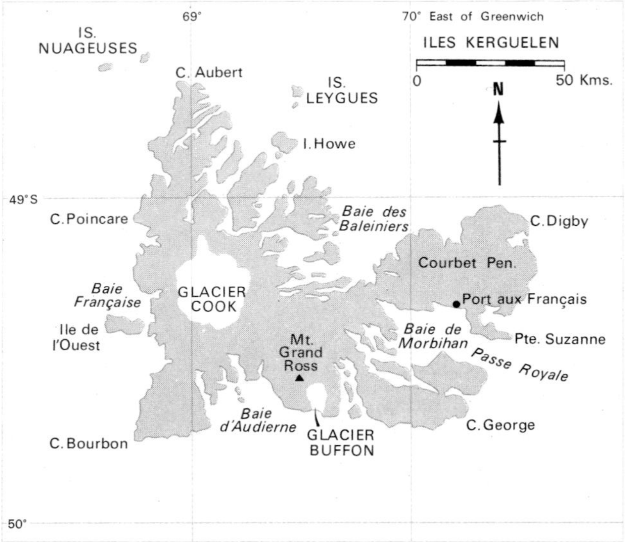

Îles Kerguelen is a deeply dissected archipelago of the southern Indian Ocean, standing athwart the Antarctic Convergence and surrounded by rich, stormy waters. Remnants of glaciers cover the peaks and uplands of the main island, but the lowlands and coast are free of snow for most of the year.

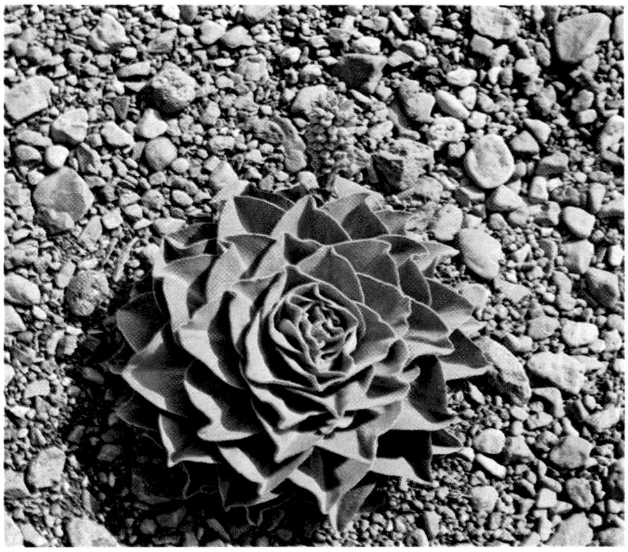

YOUNG

Kerguelen cabbage, once famous as a source of vitamin C for mariners.

Brown skua in threat display on a green, moss encrusted slope of Heard Island.

bounded by steep cliffs. Hundreds of giant petrels (probably the colonial southern species) nest on it, together with brown skuas and lesser sheathbills, and the small burrows of a diving petrel—probably the South Georgian—were found on bare ground. The southern part is a bare mound of about the same area, rising to 230 metres. Macaroni penguins form enormous colonies on Flat Island, and on the plateau, hill, and linking isthmus of Macdonald Island itself. Fur seals are plentiful on the beaches, and a few elephant seals were also recorded.

Îles Kerguelen form a compact archipelago, mostly of volcanic rocks, 550 kilometres north-northwest of Heard Island. There is one large, deeply dissected island with many peninsulas and headlands, and a number of small islets and rocks with a total area of over 7,000 square kilometres. Permanent ice fills the dead crater of Mount Grand Ross (1,960 metres) and descends to sea level down the steep Buffon Glacier. Glacier Cook, a larger ice cap, covers a high plateau and spills into narrow glacial streams down the steep west coast. Other small, depleted glacial remnants cluster about the high peaks, but long ago gave up the struggle to reach the sea. Îles Kerguelen give the impression of warming after a long period of cold, of emerging from bitter Antarctic to slightly milder Sub-antarctic conditions almost as we watch.

They stand squarely in the zone of westerly winds, almost exactly on the line of the Antarctic Convergence, ringed by a zone of confused oceanic water which results from upwelling over the sub-marine platform. A succession of depressions brings constant heavy overcast. Precipitation, more often rain than snow at sea level, falls on five days in every seven. Even in winter, monthly mean temperatures at sea level remain above freezing point; except in midwinter, snow does not lie for long on the coast. All the islands are remarkably green, with rich vegetation fertilized by the droppings of thousands of sea birds. Thirty species of birds, including twenty-eight species of sea birds, mostly in very large numbers, breed on Îles Kerguelen, and both elephant seals and fur seals are abundant on the beaches.

With their excellent sheltered harbours, Îles Kerguelen have been a centre of whaling and sealing ventures for over 150 years. By accident, or in attempts both formal and informal to farm the islands, rabbits, sheep, cats, rats, mice, reindeer, and freshwater fish have been introduced at different times, to the detriment of the native plants and animals. Now a centre for scientific research under French control, their immediate safety is assured.

Bouvetøya, discovered in 1739, is an isolated rectangular island roughly 9 kilometres by 6 kilo-

LEWIS SMITH

South Georgia has a rich variety of plant life, including many species of moss and lichen, 6 different species of fern and 17 species of flowering plant native to the island. Visitors, notably whalers and sealers, have brought many more species to the island; five are now fully naturalized and a further 22 aliens have been collected. Many are common weeds of the northern hemisphere.

Above: *Acaena adscendens*, a small burnet, forms closed colonies with moss *Tortula sp.* and grass *Festuca erecta*.

Left: The shield-fern *Polystichum mohrioides* forms dense stands, often on scree slopes and creviced rock outcrops.

Right: Bouvetøya, a tiny ice-capped island in the Atlantic sector of the Southern Ocean, stands at the southern end of the mid-Atlantic ridge.

LEWIS SMITH

European creeping buttercup, *Ranunculus repens*, introduced by whalers in the nineteenth century, growing close to the abandoned whaling station. Gritvyken, South Georgia.

LEWIS SMITH

The adder's tongue fern *Ophioglossum opacum* lives in damper situations, often on the banks of streams and waterlogged meadows. Right: *Acaena adscendens* is a tiny woody shrub with intertwining branches, growing up to 25 centimetres high on damp scree slopes and meadows. Its clumps and carpets are favourite nesting grounds of brown skuas.

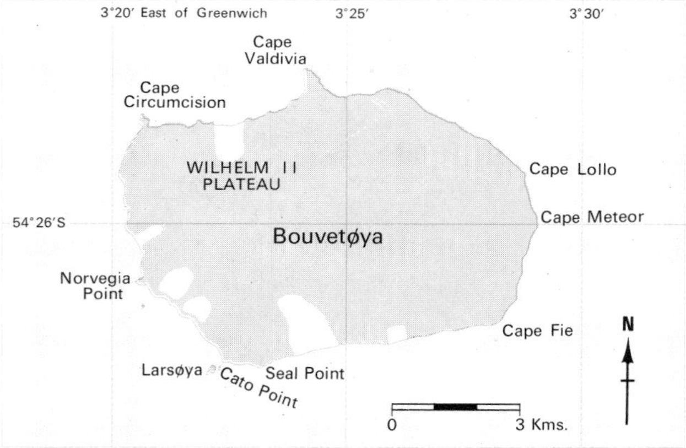

metres, rising steeply out of the Southern Ocean 3½°E. of the Greenwich meridian in 54°S. It stands at the southern end of the Mid-Atlantic Ridge, 1,600 kilometres from Dronning Maud Land and 1,850 kilometres from the South Sandwich chain. Probably a complex of volcanic cones, it rises almost symmetrically to a flat, ice-covered dome 935 metres high. Practically all of it is ice-covered. Only tiny areas of bare rock emerge from under the ice. Between February 1955 and January 1958 Bouvetøya added to itself a new platform of rock, thought by some to represent an eruption from the sea bed, by others to be no more than a rock fall, perhaps started by earth tremors. The platform, 650 metres long and up to 365 metres wide, averages 25 metres above sea level. It provides a splendid new nesting ground for Bouvetøya's chinstrap, macaroni and Adélie penguins, whose breeding had previously been restricted to Larsøya, a tiny offshore islet, and to small, highly dangerous areas of fallen rock under the cliffs of the main island. Bouvetøya also supports small populations of breeding fur seals, and of elephant, leopard and Weddell seals, who lost no time in exploring the new platform. Pintado and snow petrels, and Antarctic fulmars, are reported to breed on the cliffs; Wilson's, Antarctic, blue and giant petrels were plentiful in the air during the brief visits of research ships, but are not known to be breeding. Brown skuas live among the penguins of Larsøya.

KOOYMAN

Antarctic Birds

	Continent and continental islands	Antarctic Peninsula	Balleny Is.	Peter I Øy	South Shetland Is.	South Orkney Is.	South Sandwich Is.	Bouvetøya	South Georgia	Heard Island	Îles Kerguelen	Islands north of convergence
PENGUINS												
Emperor penguin	■	■										
King penguin							?		■	■	■	■
Adélie penguin	■	■	■	■	■	■	■	■				
Chinstrap penguin		■	■	■	■	■	■	■	■	■		
Gentoo penguin		■			■	■	■		■	■	■	■
Macaroni penguin					■	?	■	■	■	■	■	■
Rockhopper penguin									?	■	■	■
ALBATROSSES												
Wandering albatross									■		■	■
Black-browed albatross									■	■	■	■
Grey-headed albatross									■		■	■
Light-mantled sooty albatross									■	■	■	■
FULMARS												
Southern giant petrel	■	■			■	■	■		■	■		■
Northern giant petrel											■	■
Pintado petrel	■	■	■		■	■	■	■	■	■	■	■
Antarctic fulmar	■	■	?	■	■	■	■	■				
Antarctic petrel	■											
Snow petrel	■	■	■	?	■	■	■	■	■			
PRIONS												
Dove prion	?				?	■	?	?	■	■	■	■
Fulmar prion										■		■
Thin-billed prion											■	■
GADFLY PETRELS												
Great-winged petrel											■	■
White-headed petrel											■	■
Kerguelen petrel											■	■
Blue petrel											■	■
SHEARWATERS												
Grey petrel											■	■
White-chinned petrel									■		■	■
STORM PETRELS												
Wilson's storm petrel	■	■	?	?	■	■	■	?	■	■	■	■
Black-bellied storm petrel					■	■			■		■	?
Grey-backed storm petrel									■		■	■
DIVING PETRELS												
South Georgia diving petrel									■	■	■	■
Kerguelen diving petrel										■	■	■
CORMORANTS												
Antarctic cormorant		■			■	■	■		■	■		■
Kerguelen cormorant											■	■
GULLS, SKUAS, TERNS												
Dominican gull		■			■	■	■		■	■	■	■
Brown skua		■			■	■	■	■	■	■	■	■
McCormick's skua	■	■	■	■								
Antarctic tern	■	■		■	■	■	■		■	■	■	■
Kerguelen tern											■	■
LAND BIRDS												
Wattled sheathbill		■			■	■	?		■			?
Lesser sheathbill										■	■	■
South Georgia pintail									■			
Kerguelen pintail											■	
South Georgia pipit									■			

FLOOKES

Wandering albatross and chick on tussock grass nest, Bird Island, South Georgia. Wandering albatrosses nest in scattered groups on exposed hillsides where the wind helps them to take off and land safely.

Previous page: Blue-eyed cormorant, South Shetland Islands. Blue-eyed cormorants are actually brown-eyed; the blue is a brilliant ring surrounding the eye.

Left: The breeding distribution of Antarctic birds. Solid blocks indicate that breeding is confirmed, and that many birds breed where indicated. Diagonal blocks: only a few known to breed. Partial blocks: breeding only in the southern (left) or northern (right) part of the area. Blocks with query: breeding probable but unconfirmed.

Notes: King penguins have been reported in numbers on northern islands of the South Sandwich chain; they do not often assemble away from breeding colonies, so breeding is possible. Rock-hopper and macaroni penguins sometimes appear in small numbers in colonies of other species; often young birds, they may go through the motions of nesting, but seldom breed successfully until fully established. Dove prions were reported in a small breeding colony at Cape Denison in 1913, but have not been seen since. Several species of small petrel seen in the air over the cliffs of Balleny Islands, Peter I Øy, Bouvetøya, etc., almost certainly belong to breeding colonies, but the presence of nests in such inaccessible places is difficult to confirm. Antarctic terns were reported to breed near Gaussberg (Wilhelm II Land) in 1902, but have not been reported elsewhere on the continent.

LEWIS SMITH

Antarctica's most prominent animals are birds. Nearly all are sea birds; of forty-three species breeding south of the Antarctic Convergence, only five make their living entirely on land. The marine birds include seven species of penguins, twenty-four of petrels, two of cormorants, and five species of gulls, skuas and terns. The land birds include two kinds of ducks, living on the marshy flats of South Georgia and Îles Kerguelen respectively, two kinds of sheathbill—curious little wading birds which live by their wits in the intertidal zone and in penguin colonies—and a single species of songbird, the South Georgia pipit.

For a continent and several large archipelagos, forty-three breeding species is a miserably short list. The Subantarctic Falkland Islands on their own muster half as many breeding species again: temperate Britain has nearly ten times as many. Any respectable parish in Europe or North America would have a longer list—probably more songbirds alone, with a varied array of other land birds—ducks, geese, waders, marsh birds, owls, and hawks—both migrant and resident.

There are many reasons why sea birds, rather than shore, land or freshwater birds, are the main colonizers of Antarctica. Sea birds are all geared to living part of their life in water which, even in the tropics and especially in colder regions, is very much colder than body temperature. Dense water-proof plumage, thick skin and subcutaneous fat, coupled with large size and compact body form, short legs, and the least possible exposure of bare skin, help them to keep in body heat. A bird adapted for life in cold water is pre-adapted for life on cold land, and generally needs no special additional measures for coping with the Antarctic in summer. By spending as much time as possible close to the open sea, and migrating away from the continent altogether in winter, they avoid the extremes of polar climate and live comfortably on the summer riches which the sea provides. Only two Antarctic sea birds—emperor and Adélie penguins—meet the winter climate head-on.

Habitats and distribution

Land birds from temperate regions could probably survive summer on the coast of Antarctica and the maritime and peripheral islands, but would find little to eat and nowhere to live. They do, in fact, reach Antarctic islands accidentally (page 125) from time to time. But only the very few have survived to become polar residents. To show the limitations of polar habitats it is instructive to compare neighbouring islands on either side of the Convergence. The Falkland Islands (52°S.), a Subantarctic group in subpolar water throughout the year, are mild, windy, and have no permanent ice. Monthly mean temperatures remain above freezing point all the year round, and annual mean temperature is 5° to 6°C. The last glaciers melted several thousand years ago, leaving two large and many smaller islands with a varied topography of rolling hills, sheltered creeks and harbours, tidal flats, swamps and marshes, streams, moors and pasture lands. Over 145 species of flowering plants, mostly derived from neighbouring South America, provide nesting habitats and edible shoots, seeds and berries. Insects are plentiful, both in soils and in fresh water, and all the islands offer a wide range of food and habitats throughout the year for resident birds. Of sixty-five breeding species, over two-thirds feed entirely on land.

South Georgia lies 1,450 kilometres east of the Falklands and only three degrees of latitude further south. South of the Convergence, it is by contrast an Antarctic island. Surrounded by cold Antarctic Surface Water (page 36) throughout the year, its mean winter temperatures fall below freezing point and the mean for the year is 1·5°C. (page 60). South Georgia is heavily glaciated down to sea level, its lowlands blanketed from April to October. Though vegetation covers the ground completely close to sea level, and the island appears green, there are only sixteen species of flowering plants; much of the cover is moss and lichen, which neither birds nor insects find exciting. With permanent ice still present over two-thirds of the island, there is little variety of landscape—no rich estuarine mud flats, peat-encrusted moorlands or extensive pastures. Feeding habitats are poor and unvaried, and only three of the twenty-five breeding species—the South Georgia pintail and pipit, and the sheathbill—feed entirely on land.

Îles Kerguelen lie geographically north of South Georgia and on the Convergence. More varied ecologically, they have a broader avifauna, with five more breeding species. Neither snow petrels nor chinstrap penguins breed there, and Îles Kerguelen support different species of giant petrel, cormorant, sheathbill and pintail, and additional species of tern and diving petrel, prion, gadfly petrels and shearwater. The Kerguelen tern in summer feeds mostly on flying insects. The seven additional species of small petrels, all burrowing forms, probably benefit from Kerguelen's longer and warmer summers, which allow nesting to begin earlier and end later than on South Georgia. Heard Island, heavily glaciated and snow-covered, with very restricted open ground, more closely resembles South Georgia in ecology and supports only nineteen breeding species of birds. All but one—the sheathbill—feed mainly or entirely at sea.

South of the limit of pack ice, mean annual temperatures are lower, winters are longer and harder, and the vegetation cover is reduced. Tussock grass cannot grow, and only two species of flowering plants survive in sheltered, out-of-the-way corners (page 67). The South Orkney, South Shetland and South Sandwich Islands, bare and windswept, cannot support even a pipit or a pintail on land, though their sea bird colonies continue to

KOOVMAN

Blue-eyed cormorant landing, with wings stalled, feet flat on the sea. These cormorants are quite at home in the water and swim persistently below the surface in search of food.

Cape Royds, McMurdo Sound, Antarctica's southernmost Adélie penguin colony. Far southern colonies form only in places where the sea ice opens early in spring, allowing the birds to catch food for their young close at hand.

Penguins often nest in very dense colonies on the few suitable breeding grounds available to them. Macaroni penguins, Trinity Island, South Georgia.

provide food for scavenging sheathbills. But the bareness of the ground affects the variety even of sea birds. Seven species of sea birds which on South Georgia nest in or among tussock grass, are missing from islands further south. They include the grey-backed storm petrel, which nests especially in shallow tunnels among tussock roots, the South Georgian diving petrel and white-chinned petrel, which tunnel in the soil of tussock slopes, the light-mantled sooty albatross, which nests on tussock-clad cliffs, and the three large Antarctic albatrosses—the wandering, grey-headed and black-browed—which chop up tussocks to build their nests on wind-swept, grassy slopes close to the sea. King penguins often form colonies in the shelter of tussock grass. Like wandering albatrosses, king penguins keep their chicks at home throughout winter, and may not be able to breed successfully in the harder conditions south of the limit of pack ice. Other species are less particular: dove prions, Wilson's petrels, giant petrels, Antarctic cormorants and gentoo penguins breed happily among tussock grass when it is there, but as successfully on bare or moss-covered ground in the south.

Antarctica itself supports twelve breeding species of birds, all of which feed mainly at sea. Only two kinds of penguin, emperor and Adélie, are adapted for life on the continental coast. There are four species of cliff-nesting fulmars (Antarctic petrel, Antarctic fulmar, snow petrel and pintado petrel), and two smaller petrels, Wilson's storm petrels, plentiful about the continental coast, and dove prions, known only from Cape Denison, George V Land. Southern giant petrels colonize flat ground on some of the continental islands, and Antarctic terns are reported from one locality in Wilhelm II Land. McCormick's skuas keep a watchful eye on every penguin colony and every group of breeding petrels. Antarctic Peninsula, with its warmer summers and relatively longer spells of open water in spring and autumn, supports six additional species which do not breed on the continental coast. They include chinstrap penguins and a distinctive southern race of gentoo penguins (page 99), Antarctic or blue-eyed cormorants, dominican gulls, brown skuas (which in some places nest close to McCormick's skuas) and sheathbills. Two continental species—dove prions and Antarctic petrels—are not known to breed on the Peninsula coast, though they may yet be found on the many unexplored coastal cliffs and island peaks.

On both continent and Peninsula, birds tend to gather in places where, for one reason or another, sea ice disperses early in the season (page 41) and allows breeding to begin in good time. In the far south—e.g. in southern Victoria Land and McMurdo Sound—the presence of penguin colonies is a useful indication that open water is normally present from October or November onward. Not all sea bird colonies are coastal. Snow petrels breed among the high peaks of the Sør Rondane Mountains, 2,000 metres above sea level and over 300 kilometres from the coast of Dronning Maud Land, and similar colonies of snow petrels and Antarctic petrels have been reported among the coastal and inland peaks of Byrd and Victoria Lands. Usually the petrels are accompanied by small groups of McCormick's skuas, who nest among them and presumably take a small toll of adult and young birds.

THURSTON

Like most other birds, sleeping emperors tuck their bill under the wing. The chick sleeps peacefully in its blanket of feathers.

Emperor penguins with month-old chicks, Dion Islands, Marguerite Bay. This small colony, of just over a hundred birds, is one of the very few which forms on an island. Most emperors breed in colonies of several thousand birds on sea ice.

The penguins

Penguins are flightless birds, confined to the southern hemisphere and adapted for life on land and in water. Of the eighteen living species, three breed exclusively south of the Antarctic Convergence, and four breed on either side of the Convergence in Antarctic and Subantarctic zones. Penguins evolved from flying birds some 70 to 100 million years ago, probably on the temperate southern shores of Gondwanaland (page 18). Adapted originally for life in cool water, their spread into polar waters is a relatively new departure for the family. Emperor and Adélie penguins have penetrated farthest south, and show the most striking adaptations for polar life. Chinstraps, macaroni and southern gentoo penguins are efficient in the maritime

Antarctic, but do not breed on the continental coast. Rockhopper, king and northern races of gentoo penguins live on the Antarctic fringe, spanning the Convergence and seldom appearing south of their breeding islands.

All the Antarctic penguins breed colonially, usually in very large colonies numbering thousands or tens of thousands. Emperors and kings, largest of all living species, hold their single egg on their feet, dispensing with a nest. All other species build small nests of stones, bones, tufts of moss, or whatever material they can find, and lay two eggs. Among emperors, only the males incubate; in every other species the parents take turns, guarding the eggs effectively against chilling and the attentions of predatory skuas. Both parents tend the chicks, which grow rapidly during the summer. Chicks of most species are ready to leave the nest before the end of autumn, and the parents have time to fatten for a post-nuptial moult (taking three to five weeks) before winter begins. All the Antarctic species of penguins feed on plankton. Some, e.g. emperor and king, feed also on squid and fish, and may dive deep for their food.

Penguins spend about half their lives in water—far more than most sea birds—and are especially well insulated against losses of heat. They have a complete covering of feathers all over the body, excluding only the bill, feet and brood-patch. There are about twelve feathers to every square centimetre of body surface. Each feather curves to overlap its neighbours, forming a tile-like pattern which water cannot penetrate. Tufts of down (aftershafts) grow from the base of each feather, combining to form an undershirt which traps air close to the skin and helps to prevent heat losses. The skin itself is thick, and underlain by thick fat which is both a food store and additional insulation. So effective is their insulation that, at air temperatures a few degrees below freezing point, snow settles without melting on an incubating penguin. Active penguins—courting, fighting or running—cannot lose heat rapidly through their feathered surfaces, and tend readily to overheat. Adélies incubating at sub-zero temperatures seldom seem distressed by cold. When the sun shines they gape, passing air rapidly over the moist inner surface of the bill to lose heat. Feet and flippers, well supplied with surface blood vessels, can be flushed with blood—turned on like radiators—to increase heat losses during intense activity.

All penguins feed on krill and other planktonic crustaceans, and on small fish when they become plentiful. Emperor and king penguins feed also on squid and larger fish, which they probably catch in deep water. The main enemy of penguins is the leopard seal (page 148); fur seals also chase penguins in the water and occasionally on land.

Emperors bring squid and fish back from the sea, holding it in their crop and feeding it in semi-digested form to their young.

Right: Emperors in a hurry. Normally walking upright with majestic gait, emperors can also toboggan rapidly over snow, using feet and flippers to push themselves along.

Emperor penguins Largest of all living species, emperor penguins stand almost one metre high and weigh between 20 and 46 kilograms. By far the most colourful penguins, they have a blue-grey back, black cap, brilliant orange ear-patches, lilac bill plates, and lemon-yellow shirt-fronts. About two dozen colonies are known from widely scattered points about the Antarctic coast. All but two form regularly on sea ice. The emperors' breeding begins in late March and April. While all other species are heading north before the advancing winter, emperors alone move south toward their traditional colony sites, walking, swimming and tobogganing over the newly-formed fast ice. By late April most of the birds have assembled, and pairing begins. Many birds seek their partner of the previous year, wandering restlessly about the gathering, calling with braying musical cadences, listening, and responding excitedly when they hear a familiar voice. Paired birds bow and bray

repeatedly to each other, standing together or strutting solemnly with orange ear-patches flashing. Occasionally they jostle for position or squabble mildly over partners, but courtship on the whole is a leisurely, graceful affair. The first eggs—greenish white, with a chalky surface layer—appear in May at some northern colonies, June or early July in the far south. All the eggs in a colony are laid within two to four weeks of each other. A few hours after laying, the males take the egg from their partner, balance it across their toes, drape a warm fold of feathered abdominal skin over it, and settle into a crouching incubation position. Almost immediately the females head northward over the sea ice to open water, leaving their mates to incubate without respite for two months. This curious breeding routine is imposed on emperors by their environment, and by the length of time which their chicks take to reach independence. Emperors must time their breeding cycle so that the chicks reach independence in December or January, when the sea ice is dispersing and food is plentiful. Maturing in late autumn or at any other time of year would be disadvantageous and probably fatal for the young birds: so incubation must start in late May or June, and chicks must begin life in the

Emperors dive to depths of more than 250 metres, often hunting under continuous ice cover. They can stay below for more than 18 minutes between breaths.

coldest months of winter. The parents probably find it warmer to incubate on sea ice than on land, for the sea beneath the ice remains close to freezing point, while the rocks may be very much colder. With the ice edge so far away—possibly 100 kilometres or more at depths of winter—and food difficult to catch, it is more economical for one parent to incubate than for two to travel constantly back and forth across the ice relieving each other.

So the males settle to their long vigil. Having already starved for two months during courtship, they face a further two months without food before their partners return. Only a large emperor-sized bird could cope with this problem. During these four months of winter, when temperatures often fall to between −40°C. and −60°C., the incubating males live entirely on food reserves in their body, mostly on the fat which is stored in the abdomen and under the skin. They huddle tightly together in huge, sleepy masses, especially in very cold or windy weather. By reducing surface heat losses, this behaviour helps them to save energy and spin out their reserves of fat. During normal courtship and incubation emperors lose on average 120 grams weight per day, a rate which allows them to last out the four months. By the end of incubation they are down to about half their original body weight. Birds kept experimentally on their own, in similar climatic conditions, lose two or three times as much weight per day. So extravagant a rate cannot be maintained long enough to complete incubation. In the most practical terms, emperor penguins depend on each other for their success as a species.

Two months after their disappearance the females return, running and tobogganing excitedly over the sea ice to find their mates. At about the same time the chicks begin to hatch. So the female on arrival may take over an egg, or a small, piping, silver-grey chick with black head. Should the chick hatch before its mother returns, a small remnant of yolk in its stomach tides it over. In addition, the males—despite their long fast—produce small quantities of crop secretion containing fat and protein, which they feed to the chicks on demand.

Relieved of responsibility, the males leave the colony for a spell of three to four weeks at sea. The females settle to hold the chicks on their feet, feeding them on the part-digested food which they carry in their crop. For these first few weeks the chicks grow slowly. When the males return with a further load of food, growth rate accelerates. After seven or eight weeks the chicks are big enough to stand on their own feet. Growing a thick, woolly grey down, they form their own juvenile huddles, and the parents visit alternately every few days, bringing more and more food. By late October the chicks are receiving about one-third of their weight in food every second or third day, while the parents race back and forth between the colony and the sea.

Now the ice edge is starting to break back toward the colony, so that the feeding journeys become shorter. By December or early January the chicks are losing their dense grey down and acquiring a grey and yellow feathered suit—a drab version of their parents' glowing plumage. Though only a little over half-sized, they are almost ready for independence. Finally the sea ice beneath them breaks away from the land and moves north to join the pack. The young birds travel with it, swimming, feeding, and continuing to grow among the rich summer plankton. The parents, too, travel north and moult during February and March. Whether they breed again in the following season has not been determined. Theoretically at least, they have time to complete their moult, fatten quickly, and return south to begin a new breeding cycle in late March or April.

Despite their harsh environment, emperor penguins are surprisingly successful. In average-to-good seasons only one egg in every ten is lost, and about three-quarters of the parents raise a chick successfully to independence. This compares favourably with breeding success in temperate and sub-polar species of penguins, and with other polar birds. Heavy snowfall is the greatest hazard affecting eggs or chicks. Winter predation is slight; there are no skuas, but giant petrels sometimes hang about the colonies to feed on weakling chicks. The largest known colony, at Coulman Island in the Ross Sea, contains over 100,000 breeding birds, and a total Antarctic population of about 250,000 breeding birds has been calculated. Emperors seem to be maintaining their numbers, and should long continue to add welcome colour to the scene.

Adélie penguins These are smaller birds, weighing five to six kilograms in their pre-breeding fat. Less colourful than emperors, their dorsal feathers are black, tipped with blue, their shirt-fronts creamy-white. For their size, Adélie penguins are very well insulated. Individual feathers, on average 3·6 centimetres long, are only slightly shorter than emperors'. Best known of all penguins, Adélies have provided many generations of Antarctic explorers

First arrivals at the breeding colony. Adélies nest in traditional colonies, usually on the same site as in previous years, often with the same mate. Males arrive first, taking up and defending their sites as the snow cover disappears.

Adélies leap four times their own height from the sea to land on the ice foot at low tide. After a brief inspection of the ice edge they withdraw to make a rapid underwater sprint which gives them the impetus for take off.

BONE

BONE

Adélie penguins "porpoising". Travelling at speed they swim alternately over and under the water, taking a rapid breath while they are in the air.

Adélie penguins. The plumage is windproof and waterproof, forming a double layer 2 to 3 centimetres thick over the whole body. Erect crest and rolling eye are warning display.

Adélie penguin colonies always include a number of separate breeding areas, each containing hundreds of nesting pairs. Oldest and most experienced birds tend to nest in the middle of groups, where they are safest from predation by skuas.

BONE

LEWIS SMITH

Ecstatic display. Newly returned male Adélies advertise their presence by this display, with slowly flapping wings and throbbing call. It says in effect "I have a nest site and am looking for a mate".

Paired birds strengthen their bond by this reassuring display, which usually follows a disturbance in the colony.

The nest begins as a rough scrape to which both birds contribute pebbles, bones and other debris. These raise the eggs slightly above colony level and keep them dry when the ground thaws in the summer.

LEWIS SMITH

with entertainment, study material, palatable eggs and tough but tasty meat. During the past two decades long-term banding studies have yielded new information about their breeding, social organization and life expectancy.

Their colonies form on islands, beaches and headlands all round the Antarctic coast, along the western shore of Antarctic Peninsula, and on many of the maritime islands. Several colonies of over one million birds have been reported, and groups of tens or hundreds of thousands are not unusual. Adélie penguins breed between October and March. After wintering in the pack ice unencumbered by family ties, mature birds of breeding age head south toward the colonies in October, often travelling far over shifting sea ice to reach land. The nest sites, marked by scattered piles of pebbles, are usually on raised ground, exposed by winds and just beginning to emerge through the snow. First to arrive are males of several years' breeding experience, who re-occupy their old sites, to be joined within a few days by females, who return to the same area and usually to their old mate. As more and more birds enter the colonies the noise level rises; males call, cackle and croon over their nest sites, females respond enthusiastically,

LEWIS SMITH

LEWIS SMITH

Nests are spaced so that neighbours are just out of each others' reach. Squabbles often occur during nest building, but peace descends as the birds begin incubating.

and there are frequent noisy battles between neighbouring pairs over sites, partners and nesting materials. The birds scratch at the nest bowls with their powerful feet, forming a hollow in the frozen ground, and build up a shallow nest with the only materials available to them—flattened pebbles and bones. The first eggs appear early in November. Most pairs produce two, with an interval of two or three days between. All the eggs in the colony are laid within about three weeks, and parents which lose their eggs cannot replace them. Within a few hours of laying, the females return to sea, leaving the males to stand the first incubation watch.

Incubation takes about thirty-three days. The male's first watch usually lasts from seven to ten days—longer if the female has far to go to find open sea, shorter if the ice edge lies close at hand. The first watch is critical, for the male may already have starved for two weeks or more during courtship, and a further ten days without food taxes his reserves to the limit. On her return the female takes over for seven to nine days while the male fattens up at sea: thereafter the two partners alternate at intervals of two to three days. The two chicks, silvery or smoky grey, hatch almost simultaneously, and are fed from the crop of whichever

Building continues after the first egg is laid, often with pebbles stolen from nearby nests. When both eggs have appeared, the female leaves for a spell of one to two weeks at sea.

LEWIS SMITH

Chicks moult in late January and February, losing their grey down and acquiring a juvenile plumage with white chin. They achieve adult plumage with dark chin and white eye ring one year later.

parent is present at the time. The chicks are brooded closely by their parents for the first two or three weeks. Growing rapidly, they develop a thicker, more woolly grey down, and join other chicks in "crèches" or nursery groups during their third or fourth weeks, leaving both parents free to satisfy their increasing demands for food. This is perhaps their most vulnerable period, when an unseasonable fall of snow can destroy many chicks overnight.

In late December and January the colonies are visited by young, non-breeding Adélies—birds in their second and third years—who wander inquisitively, occupy and abandon nest sites, and go through the motions of courtship and nest building. They are potential recruits, many of which will settle to breed there in future years. Some behave like unruly teen-age mobs, beating respectable nest-holders and trampling on chicks. However, they are just as likely to settle quietly on unoccupied sites, fiddle contentedly with nest material, and court each other. Standing among the nursery groups of chicks, they sometimes protect them from the attentions of predatory skuas. After a few days they leave, and other young birds take their place throughout the late summer.

By late January the chicks are three-quarters grown and starting to moult. By late February most have left the colonies. Smaller and thinner than adults, with blue-grey dorsal plumage and a white chin, they fish for themselves among the drifting floes offshore. Breeding adults and non-breeding juveniles complete their own moult in February and March. By early April all the Adélies have left to spend winter in the comparative warmth of the offshore pack ice, and the colonies are deserted.

Chinstrap penguins Slightly smaller and more slender than Adélies, chinstraps are the characteristic penguins of the maritime Antarctic. Nesting like Adélies in large or small colonies, they are plentiful on the South Orkney, South Shetland and South Sandwich Islands. Along the west coast of Antarctic Peninsula they breed as far south as Bismarck Strait, especially among the islands of Palmer Archipelago. There is a small resident colony of a few hundred nests on Bouvetøya. A small but expanding group on South Georgia has increased from a handful to several thousand pairs in the last forty years. Single breeding pairs have been reported from Peter I Øy, two or three pairs from Heard Island and the Balleny Islands, usually nesting or attempting to nest among other species. Wandering individuals are often seen in late summer, usually among groups of Adélies, at points along the Antarctic coast, and are reported from many Subantarctic islands. At sea they appear far from land in the Indian and Pacific Oceans, usually along the fringes of the pack ice. Several authors have suggested that this species, once confined to the Scotia Arc region, is now extending its range. However, only at South Georgia is there positive evidence of a recent change in success of colonization, and a rapid population expansion.

Chicks remain on the nest for three to four weeks, then band together in nursery groups while both parents forage at sea.

SPAULL

Chinstrap penguin. Penguin eyes often have a star-shaped or six-sided pupil which dilates widely in poor light, for example under water, and contracts to a pin-point in the vivid light of the polar summer.

Chinstrap penguins, slightly smaller and more pugnacious than Adélies, breed only in the maritime Antarctic region. They nest on similar ground, sometimes alongside Adélie colonies and begin breeding three weeks later than Adélies. Their chicks, covered with silver-grey down, moult into a juvenile plumage with the dark chin band characteristic of the species.

YOUNG

Chinstraps are, if anything, blacker and whiter than Adélies, with a distinctive narrow band of black-tipped feathers under their chin, like the strap of a guardsman's helmet. Their feathers are shorter than Adélies' and their plumage, though equally dense, is altogether thinner. Way of life and breeding cycle are very similar to Adélies'; in summer—possibly throughout the year—they eat similar food (krill and larval fish), and often the two species nest side by side in neighbouring colonies. Chinstraps arrive to breed two to three weeks later than Adélies, and nest on rocky, sloping, boulder-strewn ground, which Adélies avoid. Courtship and incubation are similar in the two species. Chinstrap chicks are silver-grey when hatched, later dull brown and usually very dirty from the mud of the colony floor. In spite of their later start in life, the chicks moult at the same time as Adélies, and are ready for the sea by February. Juveniles in their first feathered plumage are almost indistinguishable from adults, though Dr. William Sladen reports that young birds have mottled black or dark brown on the white feathers surrounding the eye. After moulting and departure from the colonies, neither young nor old birds are seen again until the following season. Their winter movements are unknown, but they probably move to the outer edge of the pack ice and the open sea.

Gentoo penguins Gentoos are closely related to Adélies and chinstraps, and are marginally the largest and heaviest of the trio. Their distinguishing marks are a white triangular flash across the forehead from eye to eye, and brilliant coral-red bill plates. Least aggressive of the pygoscelids, they keep their distance when man approaches and are correspondingly more difficult to study. Gentoos breed on most of the islands of the Scotia Arc, on Antarctic Peninsula and its flanking islands to 65°S., on the peripheral Antarctic islands, and on the cooler tussock-clad islands of the Subantarctic zone. With so wide a latitudinal range, it is not surprising to find that northern and southern stocks of the species differ from each other. Gentoos breeding south of the 60th parallel, on the South Orkney and South Shetland Islands and Antarctic Peninsula, are relatively small, weighing on average 5·5 kilograms, with short, stubby flippers, feet and bill. These are grouped in the subspecies *ellsworthi*, named after the U.S. aviator whose expedition collected specimens while waiting for the weather to improve. Gentoos breeding on South Georgia and other northern islands are generally larger,

TICKELL

WALTON

Northern gentoo penguin feeding a half-grown chick. This subspecies breeds on South Georgia and other northern islands close to the Antarctic Convergence.

Northern gentoos breeding among tussock grass, South Georgia. Unlike most other penguins, gentoos shift their colony sites from year to year. The white bird is a rare partial albino.

Southern gentoo penguin nesting on the bare gravel beaches of the South Shetland Islands. Reddish droppings show that these birds have been feeding on krill.

weighing on average 6·2 kilograms and up to 8 kilograms or more, with longer extremities and slightly shorter plumage. Northern gentoos from the different islands probably differ further among themselves; each island group may well be found to support its own local subspecies.

In breeding and general deportment gentoos show their kinship with Adélie and chinstrap penguins. Southern gentoos of the Peninsula and maritime region often nest close to the other two species, occasionally forming mixed colonies on flat ground. They begin laying in early November, usually a little later than Adélies but earlier than chinstraps. All the eggs appear within three or four weeks, and the parents share incubation and brooding. The chicks, silvery grey above and white underneath, with the characteristic orange-sided bill of the parents, seem to take a little longer to rear, but are usually ready for the sea by late February or March. Southern gentoos take the same foods as their neighbours, suffer similar hazards from climate and predators, and are equally successful in their breeding. After moulting in March, they disappear with their newly-fledged chicks for the winter, presumably to ride the pack ice like Adélies and chinstraps.

KOOYMAN

LEWIS SMITH

Young macaroni penguin in first year plumage. Bill and crest are not yet fully developed. Macaronis attain adult plumage in their second year. Below: Macaroni penguin pair. The male, with larger, heavier bill, sits on the nest in the foreground. Macaronis typically nest on tumbled screes or cliffs, avoiding the flat ground which other species prefer.

On South Georgia and other northern islands, northern gentoos nest on raised beaches, moraines, and tussock-covered hillsides facing the sun. Unlike most penguins, they tend to move their colony sites from year to year, perhaps in defence against ticks and other ectoparasites which survive from season to season among the nesting material. In the wet climate of the westerlies, where it pays to raise a nest above ground level, some northern gentoos build monumental nests of pebbles, twigs, and massive lumps of moss or turf. Others take over the crowns of tussock grass pedestals, which stand two or three feet clear of the mud. With a little

CONROY

Rockhopper penguin. In this species the crests of yellow and black feathers sweep sideways and upwards from points on either side of the bill. Bare skin at the base of the bill may help them to keep cool in hot sunny weather.

trampling, the diverging stems form a ready-made cup for the two green-white eggs, and the birds incubate well clear of the colony floor. Inexperienced pairs build small nests at the colony edge, and lose their eggs in a sea of mud when the ground thaws.

Laying begins in early September on Îles Kerguelen and the Subantarctic islands, late October on South Georgia and Heard Island. Northern birds lay within a period of six to eight weeks, the milder climate and longer summer allowing a longer season than in the far south. Incubation and chick rearing occupy December, January and February, and the chicks complete their moult and achieve independence by early March. Adults begin their own moult, and are usually resplendent again in new feathers by early April. Then begins a curious Indian summer period, when all the adults return to the nesting grounds and begin courtship all over again. They occupy nest sites, squabble, form partnerships and stand expectantly by their nests as though about to produce eggs. The cold winds and snow flurries of early May cool their ardour, and all question of breeding is forgotten. Northern gentoos remain throughout the winter close to their breeding areas. Each winter morning, hordes of them line the beaches, swim in the surf and preen in the mild winter sunshine. Flocks of several hundred leave at a time on fishing expeditions, running the gauntlet of leopard seals, who lie in wait just offshore.

Macaroni and rockhopper penguins The smaller crested penguins of the genus *Eudyptes* are well represented in Subantarctic and temperate zones. Macaronis on average weigh 4 kilograms. Mostly black and white, they have distinctive long, drooping orange-yellow plumes growing from an arc across the centre of their forehead, and a heavy, deeply sculptured bill. Most southerly representative of their genus, they form small colonies on Îles Kerguelen and some of the Subantarctic islands, and much larger colonies on Heard Island, South Georgia, the South Sandwich and South Shetland Islands. A few pairs nest on Bouvetøya and the South Orkneys, among breeding birds of other species. The royal penguin of Macquarie Island is very closely related; some authors list it as a subspecies of the macaroni.

SPAULL

Part of a large cliff colony of macaroni penguins, high above the sea on Trinity Island, South Georgia. Macaroni penguins climb by hopping, grasping the smooth rock with claws, flippers and bill.

Heard Island rain bounces off the waterproof plumage of a king penguin. Hunted for their oil during the nineteenth century, king penguins were exterminated from Heard Island and many other breeding localities. Under protection stocks are now increasing throughout the Southern Ocean.

Rockhopper penguins weigh 2·5 to 3 kilograms. Their crests are golden-yellow eyebrows, which sweep sideways and upwards on either side of their head. They have a smaller, mahogany-brown bill and red eyes, and a well-merited reputation for ferociousness toward visiting scientists. Rockhoppers breed on almost every island of the Subantarctic zone and are plentiful on Îles Kerguelen. South of the Convergence they nest in small colonies on Heard Island, and are occasionally recorded among macaronis on South Georgia.

Both macaroni and rockhopper penguins breed on steeply sloping ground, favouring tumbled cliffs, scree slopes and steep moraines. Characteristically, they are found on exposed headlands and the weather side of islands. Undaunted by rough seas, they swim easily in tempestuous surf, cling to wave-polished rocks with strong bent claws, and climb cliffs with the agility of small monkeys. Where most penguins stroll, macaronis and rockhoppers hop, usually with flippers back and head forward. Hopping from boulder to boulder, they can climb a 200 metre scree slope in less than twenty minutes. Many nest high on the cliffs, journeying up and down two or three times each day at the height of the breeding season.

Neither species shows any particular adaptation for polar life. Macaronis nesting alongside warmly-clad Adélies and southern gentoos on the South Shetland Islands, and rockhoppers braving the summer on Heard Island, wear plumage no thicker than that of other crested penguins on warm-temperate shores of New Zealand. They probably benefit by nesting among boulders and rough rock outcrops which, sunny and sheltered from wind, are the warmest sites to be found on Antarctic islands. Both spend the least possible time ashore. On Heard Island, where the species has been studied most fully, macaronis arrive at the colonies in late October and early November, and begin laying about 10 November. They lay two eggs; the first, usually the smaller by twenty to fifty per cent, is invariably thrown out of the nest or broken, and only the second is reared. Macaronis incubate for thirty-five to thirty-seven days. The parents take turns, usually with the female incubating first. The chicks, dark grey dorsally and white on the underside, form crèches in late January, moult into juvenile plumage about a month later, and leave the colonies by mid-March. Adults return en masse to the colonies in late March to moult, departing finally in mid-to-late April. Rockhoppers follow a similar time table, starting and ending their cycle two to three weeks later than macaronis. For the remaining six months of the year, from late April to late October, they disappear altogether from the breeding grounds, probably spending their entire winter at sea.

King penguins Kings belong to the same genus as emperors; they weigh on average 15 kilograms, up to 21 kilograms when in pre-moult or pre-breeding fat, and stand about 80 centimetres high. The dorsal feathers are tipped with dark powder-blue, shoulders and nape are silver-grey, and the comma-shaped ear patches, broad pink bill plates, and golden breast feathers are, if anything, more brilliant than the emperors'. Kings breed throughout the Subantarctic and on the peripheral

Antarctic islands of Kerguelen, Heard and South Georgia, forming colonies—usually of several thousand pairs—on tussock flats with large snow banks nearby. Formerly killed in thousands by oil hunters, their stocks are now recovering slowly in some of their old haunts.

King penguins share with emperors the problem of raising a large, slowly-growing chick in an environment where food is scarce for half the year, and have solved the problem in an altogether original way. Both adults and chicks are present on the colonies throughout the year. Laying begins in November. First birds to lay are those which lost their egg or chick in the previous season. They form pairs in the driest, most favourable areas of the colonies, elbowing out of the way all the fully-grown chicks from the previous season. Males take the first incubation watch, holding the single, large egg on their feet for about fifteen days. Thereafter the parents alternate at four-to-five day intervals, the relieved partner disappearing from the colony to feed at sea. Incubation takes fifty-four days. The first chicks hatch in mid-January, and are fed to full capacity by both parents from the moment of hatching. Covered first in fine grey-brown down, later in a dense brown "wool", they grow very rapidly, reaching a weight of 10 to 12 kilograms by April. Meanwhile, successful breeders of the previous season have been fattening their chicks, allowing them to moult and reach independence. After a rapid moult of their own, they have laid again in January and February, and by April have small but rapidly-growing new chicks on their feet.

In late April and May food becomes scarce in surface waters, and the behaviour of the birds changes. All courtship, incubation and brooding ceases. Eggs and small chicks are abandoned, and any chicks of less than 4 to 5 kilograms die of exposure during the first winter storms. Most adults leave the colonies for the sea. The large chicks band together in crèches, packing tightly together for warmth. Parents visit the colonies to feed their chicks at intervals of five or six weeks; thus any individual chick receives food only every second or third week. The chicks lose weight slowly. Those which began the winter at 10 to 12 kilograms fall to 6 to 8 kilograms. Any which began at less than 7 kilograms fall to 3 or 4 kilograms, and usually die. There is a small, steady mortality throughout the winter.

Then in October the plankton returns to surface waters and fishing becomes easier. The parents begin to visit more frequently, and the weight of

surviving chicks increases slowly. By November the largest survivors are back to full weight and starting to moult, and the remainder moult between December and February. Parents which lost eggs or chicks early in winter fatten and moult, ready to become the early breeders of November. Successful parents, the early breeders of last season, become the late breeders of this season. Last season's late breeders, whose chicks have still to complete their moult in January or later, may themselves moult and pair again, but are very unlikely to succeed in breeding. They become the bulk of the failed breeders, ready for an early start in the following November. Thus king penguins stand a sporting chance of raising two chicks in each period of three years, or of one every second year if their late breeding fails.

Colony of king penguins on glacial moraine, Salisbury Plain, South Georgia. King penguins, like emperors, incubate without a nest, holding the single egg on their feet.

King penguin chicks in late summer, South Georgia. The larger chicks, hatched two to three months earlier than the smaller ones, stand a better chance of surviving the oncoming winter.

LEWIS SMITH

The petrels

Petrels are a large, diverse order of sea birds, of ancient lineage and probably distantly related to penguins. All are flying birds, with dense plumage, webbed feet, a hooked bill, and curious tubular nostrils. Most feed on plankton, which they catch in a variety of ways. A few also scavenge on land, though many petrels find walking difficult and settle on land only at the breeding sites. All lay a single egg, and take longer than most other birds to incubate and rear their chicks. Banding studies show that many petrels spend several non-breeding years wandering, then return to their natal colony to breed. They live long—some of the larger ones for over fifty years. Like many other sea birds they have large nasal glands, in which they desalinate sea water to provide themselves with fresh water.

The four families of petrels are all represented in the Antarctic region, by a diversity of species and very generous numbers. The Diomedeidae include the albatrosses and mollymawks, large, heavy birds with long slender wings, which spend their lives soaring and gliding over oceanic waters, and feed by settling on the water and dipping for surface particles. The Procellariidae, a large family of many talents, include the fulmars, prions, gadfly petrels and shearwaters. Giant petrels, largest of the family, are heavy gliders who soar and swoop with the albatrosses. Smallest of the procellariids are the fluttering, delicate prions, which swirl in clouds over the ocean surface like grey-blue leaves. Even smaller are the diving petrels (Pelecanoididae), a family of thrush-sized petrels who dive, with tiny wings whirring, through wave-crests in search of food. Smallest and lightest of all are the storm petrels (Hydrobatidae), whose trick of pattering over the water on black webbed feet suggested St Peter to early mariners, and gave petrels their name.

Albatrosses and mollymawks These are the largest and most easily recognized of the petrels. *Wandering albatrosses,* largest of all flying sea birds, weigh up to 10 kilograms. Their wings, almost incredibly long and narrow, span between 2·5 and 3 metres; only the royal albatross, a closely-related species of the Subantarctic zone, matches wanderers in size and magnificence. Wandering albatrosses breed on South Georgia and Îles Kerguelen in Antarctic waters, and on every sizeable island of the Subantarctic zone. They nest on windswept hill-sides, in sight of each other but several hundred metres apart, with ample clear space for landing

TICKELL

and take-off into wind. The nests, untidy mounds of freshly cut grass and hay, are surrounded by short turf and ringed by the longer tussock grasses. Banding studies, notably by Lancelot Tickell on South Georgia, have shown that wanderers banded in the nest as chicks spend several years at sea before attempting to breed. After about their third year, they return each spring to their natal island, and in the next few summers go through an elaborate courtship ritual of dancing and nest-building with other young and unattached birds. No eggs result, but the courtship brings pairs together and helps to form life-long pair-bonds.

Mature birds of breeding age return to nest sites on South Georgia in early November, females in late November and December. Two to three weeks of courtship and desultory nest-building renew the

Courtship dance for young wandering albatrosses. From their third or fourth year onward juveniles return each season to take part in courtship rituals with other young birds. They seldom breed before their sixth or seventh years.

The albatrosses, largest Antarctic flying birds. Wandering albatrosses, enormous birds with long narrow wings and a pink bill, vary widely in colour according to age. Fledgelings are uniformly dark brown, with white face and throat. In their second, third and subsequent years, the body and inner wings become progressively paler, with many white feathers appearing among the brown. Mature birds are almost entirely white, with black wing tips and trailing edge of wing, sometimes with a dark cap and dark tail feathers. Females tend to be darker than males of the same age, and seldom reach the all-white stage. The smaller albatrosses (or mollymawks) are only half to one quarter the weight of wandering albatrosses, with relatively broader wings.

Wandering albatross
Black-browed albatross
Grey-headed albatross
Light-mantled sooty albatross
RIGNALL

TICKELL

Grey-headed albatrosses nest in cliff top or hillside colonies. Each nest is a pillar of plant debris to which both adults and chicks contribute. Chicks are fed on the nest by their parents, who bring in part-digested fish and squid. Hatched in January, they are ready to fly by April or May.

TICKELL

Black-browed albatrosses in a colony overlooking the sea. Albatrosses rely on strong winds to help them become airborne. Their nests face the wind, allowing immediate take-off and landing.

old partnership and re-establish the importance of the nest site. After both partners have returned for short spells to sea, the female begins a more concentrated bout of nest-building, and, usually between mid-December and mid-January, lays a single white egg of half a kilogram weight. The parents incubate in turns for eighty days of late summer and autumn, then brood and feed the chick for a further month to five weeks. In May, as winter weather sets in, the chick is left on its own for the first time, surveying the world blandly from its untidy, snow-encrusted nest pile. No predators molest it. Swathed in thick woolly down and well upholstered with fat, it has nothing to do but sit there and grow.

Unlike king penguins (page 104), wandering albatrosses return every few days to the nest, bringing their large offspring a regular supply of fish and squid. Perhaps flight allows them to forage more widely than penguins, or see their prey from further away. They seem to hunt mostly within easy flying distance of the island; only persistent bad weather breaks the pattern of their visits. So the weight of the chick and its rate of growth are maintained throughout the winter. From July or August until early November the chick far outweighs its parents. Then it begins to slim, and change from dense woolly down to a dull, dun-black and white juvenile plumage. It strengthens its wing and leg muscles by flapping and walking about the nest site. After final feeds from its parents it spreads its wings, and a new ocean wanderer is launched.

The parents, too, desert the nest site. Too late to begin breeding immediately, they wander unencumbered for a whole season, wintering off the coasts of South Africa or Australia in the company of younger birds. In the following November they return to breed again, usually in the same nest and with the same partner. Just over half the breeding population of wanderers breed, therefore, in any one year. The total population is estimated by Tickell to be 20,000 birds. At least a proportion of them may live for more than fifty years.

The smaller albatrosses, sometimes called mollymawks, weigh 2 to 4 kilograms and have wing spans of just over two metres. Three species, two mainly black and white and the third smoky-brown, nest within the Antarctic region. All manage to confine their breeding cycles to a single season, launching their chicks before the onset of winter. Throughout their wide pelagic range all three species feed on fish, squid and euphausiid crustaceans.

Nestling light-mantled sooty albatross. After the first month albatross chicks are left alone while both parents forage at sea. They defend themselves effectively against predators by ejecting oil from their crop. Though some die of cold and exposure, most survive their nestling period.

Grey-headed and *black-browed albatrosses* are white-bodied birds with mainly black wings; the under-surface of the wings has a broad white longitudinal band. Grey-headed albatrosses have a distinctive blue-grey head and back, and a dark grey-brown bill with bright yellow culminicorn or central plate and a narrow line of yellow on the mandibles. Black-browed albatrosses are similar, but with a white head and back, all-yellow bill, and a dark line of plumage above the eyes which gives them an appearance of profound, concentrated thought. The two species breed side by side in colonies of several hundreds or thousands, mostly on Sub-antarctic islands. In the Antarctic they are well represented on South Georgia and Iles Kerguelen, and black-browed albatrosses alone nest on Heard Island.

Colonies form among tussock grass, often on the tops and tumbling flanks of weathered sea cliffs. The nests are tall cylinders of trodden turf and mud, some standing half a metre or more above ground level, with a shallow bowl on the summit. Parents add to the nests during a noisy ritual courtship dance, which occurs in October and early November, and while away the tedious seventy or more days of incubation by plastering layers of additional mud onto their nest cylinders.

Light-mantled sooty albatrosses are exquisite small albatrosses with dark chocolate-brown head, wings and tail. The bill is dark brown, with a smart fillet of blue, and the dark eye is ringed almost completely with a narrow circle of white, which gives the birds a curiously knowledgeable air. They breed in isolated pairs on narrow cliff edges, often but not always overlooking the sea. Usually they require vegetation—tussock grass, moss, or small shrubs—to build into the nest pedestal; the most favoured sites are draped with tussock grass, protected by overhang from rain and falling debris, and offer a splendid view over the sea or surrounding country. Never so numerous as grey-headed and black-browed albatrosses, they breed on South Georgia, Heard Island, Îles Kerguelen and most of the Sub-antarctic islands, laying throughout their range in late October and November.

Chicks of all three species hatch in late December and January, and spend some four months in the nest. Brooded for the first month, they are later left, sitting quietly like woolly toys on their pedestals, while both parents forage at sea. If molested by skuas, inquisitive scientists or other possible predators, they draw themselves to full height and, opening their bill, eject a stream of red oil containing fragments of squid and fish—an effective deterrent, which usually ensures that they are left strictly alone. The chicks fledge in April and leave the colonies in late April or May. Neither parents nor young birds are seen about the colonies in winter. All three species disperse widely over the southern oceans, from subtropical waters to the edge of the pack ice and beyond. Light-mantled sooty albatrosses are frequent visitors to the Ross Sea and the Antarctic coast. While sooty and black-browed albatrosses are believed to nest annually, there is some evidence from banding that grey-headed albatrosses breed only every second year.

Fulmars The fulmars are large to medium-sized petrels with a strong bill, heavy body and short, square-cut tail. In the winds of the roaring forties they have a characteristic gliding and flapping flight, with notable aerobatic skill. Generally ocean scavengers, some have developed filtering bars on their upper mandible, which help them to feed on plankton.

Largest of the fulmars is the *giant petrel*. Up to a few years ago this was regarded as a single, rather variable species, but it is now clear that there are two forms distinct enough to be called separate

TICKELL

BUDD

Southern giant petrels nest on islands fringing the continent and on maritime Antarctic islands. At the northern edge of the range many have dark plumage, like this one nesting on South Georgia.

Above, right: Giant petrel feeding on a dead elephant seal, Heard Island. Giant petrels are scavengers, cleaning up carcases of penguins and seals. Though barely able to walk upright on land, they lurch with wings spread and legs straddled, and fight noisily to keep others away from their prey.

Giant petrels of intermediate colour nest throughout the southern region. The single large egg, laid in October or November, is incubated for two months in an untidy nest of gravel and pebbles. The chick, seen here on the point of hatching, grows to full adult size in only four months.

BURLEY

species. The southern giant petrel breeds in colonies on fringing islands of continental Antarctica and the Peninsula, Scotia Arc, Heard Island, and Macquarie and the Falkland Islands in the Subantarctic. Mostly dull grey or black mottled birds, with greenish bill, individuals of this species nesting in the far south tend to be pale—even white—while those of the northern Antarctic islands tend to be darker. They breed colonially, usually nesting from late September onward. The northern giant petrel nests almost entirely in the Subantarctic, becoming an Antarctic breeder only on Îles Kerguelen. This is a uniformly dark species, but always has a distinctive pale face and pink bill, and breeds in solitary nests, usually in early spring. Both species feed mainly at sea, swooping like albatrosses in search of food, settling with wings part-folded, and gulping down fish, krill, squid, and large numbers of smaller birds impartially. They are also land scavengers, using their powerful bills to tear open seal carcases and fallen penguins.

Giant petrels build untidy nests of stones, vegetation and debris on open ground, usually where the wind allows them a quick take-off. Birds of the southern species, nesting from September onward, incubate for about sixty days, brood their chicks for three to four weeks, and fatten them to adult weight in seven to eight weeks. The chicks reach independence three and a half to four months from hatching, and adults probably nest each year. Curiously timid in comparison with penguins and albatrosses, giant petrels have not been easy to study as breeding birds. They can, however, be approached at the nest late in incubation or while brooding small chicks, and many thousands of breeding adults, and chicks of known age, have been banded throughout the Antarctic and Subantarctic. Like albatrosses, giant petrels scatter widely across the southern ocean on leaving the nest as fledglings. Dr. William Sladen and his colleagues, who have been responsible for long-term banding studies of this and other species, report that one newly-fledged juvenile flew from the Windmill Islands, on the Wilkes Land coast, to Tasman Bay in New Zealand—a distance of over 4,000 kilometres in two weeks. Other fledglings banded at the same colony were recovered within a few weeks in New Zealand, Chile, Peru, Argentina and southern Australia. Again like albatrosses, giant petrels spend several years wandering before returning to breed at or close to their natal colony. Few seem to attempt breeding before their seventh year, and breeding success is low before the tenth or eleventh years.

The smaller fulmars of Antarctic waters range in weight from 300 to 900 grams, and differ physically in build, colour patterns and flight. Ecologically they are very similar. All nest in open cavities or sheltered ledges, often on cliffs or steep, scree-covered slopes. They are completely at home on the bare rocks of the polar continent, and tend to select bare ground even when nesting on the northern islands. They breed colonially, sometimes in mixed groups, occasionally squabbling over nest sites. Often they feed together in large or small mixed flocks, though each species also has its own feeding areas which partly or completely exclude others. Euphausiids, small fish and squid form the bulk of their diet. Nesting mortality is high, especially in

The fulmars. Chequered pintado petrels (cape pigeons) and all-white snow petrels are unmistakable. Antarctic petrels have a clean, distinctive pattern of brown and white plumage, with dark-tipped tail. Antarctic fulmars are grey and gull-like, though unlikely to be confused with any gull of the southern hemisphere. Giant petrels are fulmars of mollymawk size, with very variable colour patterns. All have an enormously heavy bill, pinkish in the northern species, yellow-green in the southern.

EVERSON

Pintado ("painted") petrels breed on many Antarctic and sub-Antarctic islands and roam widely across the Southern Ocean in feeding flocks. Their prominent, chequered plumage pattern distinguishes them from all other petrels. They feed on krill and other small animals of the plankton. They often accompany whalers and sealers to feed on scraps from the carcases.

PRÉVOST

Antarctic fulmar with chick at Point Géologie, Terre Adélie. Closely akin to the fulmars of the North Atlantic ocean, Antarctic fulmars nest on the cliffs of mainland Antarctica and neighbouring islands.

Snow petrels are pure white with black bill, eyes and feet. Dark down under the feathers, invisible at the surface, helps them to keep warm in the cold air of Antarctica. They fly and feed among pack ice and breed only within the pack ice zone.

BURNIP

the far south where unseasonable weather may destroy more than half the annual crop of eggs or nestlings.

Antarctic fulmars, largest of the group, weigh 600 to 900 grams. Elegant grey and white birds, they have long, gull-like wings but an unmistakably chunky fulmarine head, body and tail. They breed on the continent, on every island group within the limit of pack ice, and on Bouvetøya. Continental birds return to their cliff nest sites in October, shovel away the winter accumulation of snow with bills and feet, and court actively for two months before laying. Females disappear for a few days to feed at sea, then return to lay early in December. All the eggs in a colony of several dozen nests are laid within a few days of each other. Both parents incubate, changing watches at intervals of three to four days, for a total incubation period of forty-five to forty-seven days. Hatching in mid-January, the chicks grow rapidly, reaching and exceeding the weight of their parents by mid-February. They fly in their eighth week, and the colonies are deserted by the end of March.

Pintado petrels are dark brown and white, with a chequered pattern on the upper surface of their wings. Their breeding range extends from continental islands to some of the temperate islands south of New Zealand, and wandering flocks are occasionally met over subtropical waters off Africa, Australia and South America. Pintado petrels—for long known to whalers as "Cape pigeons"—are

The prions. Small, fluttering blue-grey petrels, usually seen in very large flocks and almost impossible to distinguish on the wing. Note the broad bill of the dove prion, the strong, chunky bill of the fulmar prion, and the slender, delicate bill of the thin-billed prion.

especially prominent as scavengers. Huge flocks of them follow whaling and sealing ships as they travel through Antarctic waters, and feed busily on scraps of meat and blubber during flensing operations. Their bill, like that of the prions (above), has lamellae which help them to pick up small particles of food. They nest in colonies on cliffs and scree slopes, following a breeding routine which differs only in detail from the routine of Antarctic fulmars which is described above.

Antarctic petrels are striking brown and white birds: head, neck, back, tail and leading edges of the wings are chocolate brown, and the rest of the body is pale tan or white. They breed only on the coast of Antarctica, nesting typically on sea cliffs in colonies of up to a million pairs. Only eight or nine breeding areas are known. On the Windmill Islands, where a small colony was studied by Dr. M. N. Orton, they take up their breeding stations early in October before other petrels have returned, occupying the sunniest and most sheltered breeding sites. About a week later many are ousted by Antarctic fulmars, who are larger and heavier. Their breeding is similar to antarctic fulmars', and the chicks reach independence slightly earlier, usually by the end of February. Antarctic and pintado petrels were seen near the Windmill Islands in July, suggesting that at least a proportion of these species winter close to the coast and inshore pack ice.

Snow petrels, smallest of the fulmarine petrels, weigh only 300 to 500 grams. They are pure white, with black eyes, bill and legs. Their underdown, invisible at the surface, is also black, perhaps helping to warm them by absorbing radiation which has penetrated their snowy outer plumage. They breed throughout the continental and maritime regions, usually in small groups on cliffs, scree slopes and exposed hill tops. Several inland colonies are reported, one of them over 300 kilometres from the nearest coast. As the French biologist Jean Prévost has shown, snow petrels from Pointe Géologie, in Adélie Land, are larger and heavier than those from most other breeding localities, and should probably be regarded as a separate subspecies. They feed mainly about the pack ice, fluttering over leads and settling briefly on the water.

The prions or whalebirds are small, smoky-blue petrels with large wings, white underparts, and a black tip to their tail feathers. Bill and feet are blue. The three Antarctic breeding species are closely similar, and easily confused with Subantarctic forms which sometimes feed in Antarctic waters. All weigh between 100 and 200 grams. Prions feed by hydroplaning over the water with bill immersed and feet paddling. Their bills contain tiny comb-like plates which trap food particles, and the skin of the mandibles forms a pouch which shoots water out from the sides. Thus the birds pick up both particles and water, hold the particles momentarily on the plates, and spit out the water. They also plunge and peck, feeding on crustacea, small squid and other planktonic animals; different shapes of bill and modes of feeding suggest that the several species of prion take different foods, or in part divide the plankton between them. Feeding by day, they

CONROY

Dove prions breed underground, digging long tunnels into banks of moss, and laying their single egg in a subterranean chamber. This helps them to avoid extremes of cold and heat and keeps them safe from the attentions of skuas. They fly about the breeding colonies only at night.

Gadfly petrels. Blue petrels are remarkably prion-like, distinct only in their white-tipped tail and dark forehead. The remainder are darker birds, with fluttering flight. White-headed petrels, with white face and pale body, are easiest to distinguish. Kerguelen and great-winged petrels are uniformly grey brown and it is hard to tell them apart.

enter and leave their nests only at night. Almost defenceless on land or in the air, their size makes them easy prey for skuas, and they cannot afford to be caught over land in daylight. They nest underground, in cracks and fissures or in deep burrows under soil and vegetation.

Dove prions breed on the South Orkney and South Sandwich Islands, South Georgia, Heard Island, Îles Kerguelen, and the Subantarctic islands of Macquarie and Auckland. Lancelot Tickell's statistical studies have shown that dove prions of Îles Kerguelen differ subspecifically from those of Heard Island and the Scotia Arc, the Subantarctic stocks forming a third, eastern subspecies. A small breeding colony was reported in 1913 at Cape Denison, on the Antarctic continent, but has not been seen since. *Fulmar prions* breed on Heard Island, mostly on inaccessible ledges on sea cliffs, and in similar situations on the New Zealand Subantarctic islands. *Thin-billed prions* breed in small numbers on Kerguelen, and on the Falkland Islands in the Subantarctic zone.

The gadfly petrels are mostly Subantarctic, temperate and tropical birds, small but heavily built, with long wings and a distinctive swooping flight. Their bills, heavy and sharp, seem right for catching and cutting up squid, which they hunt mostly at night. Four species enter the Antarctic zone, breeding at Îles Kerguelen and feeding to varying degrees in Antarctic waters. Little is known of their breeding biology. *Blue petrels* are gadfly petrels which seem to be trying hard to turn

RIGNALL

White-chinned petrels earned their nickname "shoemaker" from their habit of sitting underground and tapping. They nest in deep burrows and are seldom seen on the surface during the day. Shoemaker burrows are dug among the roots of tussock grass. This species nests only on islands north of the limit of pack ice where tussock grass flourishes. In spring they tidy their burrows, scraping out remnants of last year's nesting.

WALTON

RIGNALL

Shearwaters. Large petrels with distinctive soaring flight, the shearwaters dabble, plunge and dive for their food. Grey petrels have a dark blue-grey body with slightly paler underparts and heavy dark bill. White-chinned petrels are uniformly darker with a clean white chin.

into prions; in colour, shape, nocturnal habit and many other ways they have converged toward the prion pattern. They nest in tunnels, which they dig in soft ground under vegetation, laying in late October and early November. Intensely social, the birds leave their colonies en masse shortly before laying, and re-visit in hordes at intervals during the winter. *Great-winged, white-headed* and *Kerguelen petrels* are three very similar species of brown gadfly petrels. Great-winged petrels, dark brown with black wings and tail, are winter breeders, nesting in May and early June on Îles Kerguelen. The chicks are ready for independence in November and December. Kerguelen petrels lay in late October, releasing their chicks in late January and early February. White-headed petrels lay in late December, and the chicks remain in the nest until May or early June.

Shearwaters are medium-to-large petrels with long wings, powerful feet (which they use in burrowing) and a long, sharp bill. They feed by dabbling at the surface, especially for squid, but some species also dive and follow their prey under water. Two species nest south of the Convergence. *Grey petrels* have brown wings and tail, and a delicate blue-grey head and chest. Heavy birds, weighing about one kilogram, they nest on many Subantarctic islands and on Îles Kerguelen. Eggs are laid in deep tunnels during March, and the chicks reach independence in late winter—August or early September. *White-chinned petrels* breed in the Subantarctic and on Kerguelen and South Georgia. Slightly heavier than grey petrels, they have a yellow bill and distinctive pale chin. Their eggs appear in November, and the chicks in April.

Diving and storm petrels The diving petrels are small, compact birds, weighing 120 to 220 grams, usually dark blue-grey or black, with stubby wings which drive them effectively through air or water. Typically, they fly low, dodging through the crests of waves to snap up their planktonic food. On land they burrow in large colonies. The *South Georgian diving petrel* nests on South Georgia, Heard Island, Îles Kerguelen and Macquarie Island, and feeds far south toward the edge of the pack ice. They dig their nest burrows in bare soil or sand, sometimes high on hilltops, and lay in early December. The *Kerguelen diving petrel,* which breeds only on Îles Kerguelen, Heard Island, Îles Crozet and the Auckland Islands, is a slightly larger bird closely akin to (perhaps even a subspecies of) *Pelecanoides urinatrix*, a species widespread in the southern cool temperate region. On Heard Island and Îles Kerguelen it burrows in grassy slopes, often close to the sea, and lays in early-to-mid December.

The storm petrels are tiny birds weighing 30 to 60 grams, with relatively enormous wings, long legs, and delicate black feet. They flutter over the water in small, busy flocks, hovering against the wind, pecking frantically at the surface with small black bills, occasionally settling in calm water to dabble at leisure. They nest in colonies, usually in short burrows, cracks or crevices. Like most other small petrels, they seldom appear over land during the day, but come ashore in late evening to court, take over incubation or feed their chicks, and leave again at first morning light. *Wilson's storm petrel* has a very wide breeding distribution, nesting on the Antarctic continental coast and on practically every island in the Antarctic region. Several subspecies have been proposed, and it seems very likely that separate breeding stocks can be distinguished within the species as a whole. Wilson's petrels migrate north after breeding, and spend April to November north of the equator. *Black-bellied* and *grey-backed storm petrels* are more restricted in their breeding, the former to the Scotia Arc, Îles Kerguelen and on Subantarctic islands south of New Zealand, the latter to the Subantarctic, Îles Kerguelen and South Georgia only. South of the Convergence, all nest in November and December, completing their breeding cycle in April.

Coastal and land birds

Apart from its many millions of oceanic birds, the Antarctic region supports smaller numbers of coastal birds—cormorants, gulls, terns and skuas—which feed inshore, in the intertidal zone, or on penguin colonies. Only a few species have moved south to take up this way of life, and their numbers remain small because, instead of an unlimited ocean, they have only restricted waters or narrow stretches of shore to feed on. True land birds—those which feed entirely along the shore or inland—are even more restricted in choice of habitat.

The cormorants or shags of the far south exploit the niche of seaside cormorants all the world over. They nest close to the sea, often near sheltered water, dive from rocks, and swim persistently below the surface—often at depths of several metres—in search of fish, squid, and bottom-living worms, cockles and other molluscs, and browsing crustaceans. They build nests of seaweed, feathers and rubbish of various kinds, cementing them together with guano or droppings. Southern cormorants lay from two or three eggs in the far south to five in the north, mostly in November and December, and the chicks reach independence in February and March. Southern colonies form where the sea ice opens early in spring, though temporary freezing-over may send the birds flying long distances in search of open water and food.

Of the two Antarctic species, one, the *blue-eyed cormorant*, breeds at small, scattered colonies in the Scotia Arc, and south along Antarctic Peninsula to the southern end of Adelaide Island. The same

RIGNALL

Diving (left) and storm petrels. Smallest of the petrels, these two forms take their food from the ocean surface in different ways. Diving petrels plunge like bullets into the waves, emerging with wings whirring after brief diving forays. Storm petrels, only one quarter as heavy, flutter like leaves over the ocean surface, dipping and dabbling for their food.

The Antarctic or blue-eyed cormorant is one of two species of cormorant breeding in the Antarctic region. It nests in South America and the Scotia Arc.

RIGNALL

Wilson's storm petrels, like other small petrels, patter over the surface of the water with wings outstretched. They feed on tiny animals and small particles of the plankton, less than 1 centimetre long. Most of their food is picked from the surface.

LEWIS SMITH

RIGNALL

LEWIS SMITH

Dominican gulls are smart black and white birds similar to greater black-backed gulls of northern latitudes. Eggs and chicks are cryptically coloured to match the background of moss, lichens and pebbles of the nesting sites.

The Kerguelen cormorant is very similar to the Antarctic cormorant but does not have the white bands on its wings. It breeds on Îles Kerguelen.

species is found in southern South America and on the Falkland Islands, and a distinctive subspecies *(nivalis)* nests on Heard Island. Local breeding stocks in South America, South Georgia and the southern Scotia Arc differ slightly from each other, and are regarded by some authors as distinct subspecies. Blue-eyed cormorants are smart birds, with blue-black dorsal plumage, black-crested head, and a white throat, chest and abdomen. White bars appear on the dorsal surface of the wings, and a patch of white feathers on the back while chicks are in the nest. The face bears bright yellow caruncles or knobs, and the eyes are ringed with a narrow circle of Cambridge-blue skin. A second Antarctic cormorant, called by some authors the Kerguelen cormorant, nests on Îles Kerguelen. This is a local representative—best regarded as a subspecies *(verrucosus)* of another South American species, the *king cormorant,* which has spread eastward to nest on several Subantarctic islands. It lacks the transient white wing bars and tail patch of the blue-eyed cormorant, and has rather less white on its cheeks, but is otherwise confusingly similar.

Only one species of true gull, the *dominican gull,* breeds in the Antarctic region. A large white gull with black upper wing surface and mantle, it is common in summer throughout the Scotia Arc, along the Peninsula south to Marguerite Bay, and on Heard Island and Îles Kerguelen. The same species, with little or no local variation, breeds also throughout the Subantarctic and in South America, South Africa and New Zealand. In the south dominican gulls nest alone or in small communities, usually on raised beaches or among rocks of the shore, building with stones, bones, grass and other local materials. Two or three eggs are laid in November or December, and incubated by the parents alternately for about twenty-seven days. The chicks leave their nest two or three days after hatching and fly in six to seven weeks. First-, second- and third-year plumages are distinctive, with grey-brown barring. Fourth-year birds look very much like adults, but may retain a few brown feathers on the head and neck. The birds probably breed in their fifth or later years. Dominican gulls feed along the shore, tweaking limpets from rocks in the intertidal zone, turning over seaweeds, scouring the beaches for washed-up plankton and storm debris. Though carefully avoiding direct competition with skuas (see below), they are sometimes found scavenging on penguin colonies, cleaning up seal carcases, and invading the rubbish dumps of

YOUNG

Brown skua eggs in a nest of matching mosses and lichens. Deception Island, South Shetlands.

A fledgling Dominican gull with tattered remnants of down crouches among the vegetation of a maritime Antarctic island. Second year birds wear similar but paler plumage, third year birds more closely resemble adults. Full adult plumage is achieved in the fourth year.

BONE

scientific bases. Resident throughout the year on northern islands, they desert southern stations along the Peninsula in autumn and return shortly before breeding in spring.

Skuas are large, gull-like birds with dull brown plumage, prominent pale under-wing bars which appear in display and flight (page 79), and grey-black bill and feet. The two Antarctic species, similar in habit and appearance, replace each other in the maritime region. *McCormick's skua*, the southern form, breeds on the continent and Antarctic peninsula, and also on the Balleny Islands and Peter I Øy. *Brown skuas*, slightly larger and darker, with notably longer legs and bill, breed throughout the Scotia Arc and on all the more northerly Antarctic and Subantarctic islands. McCormick's skuas of the Peninsula tend to be darker than those of the continent, and the degree to which these two confusing species overlap in the South Shetlands and Peninsula area has never been satisfactorily determined. Both nest in November and December in the south and about a month earlier on the northern Antarctic islands, forming a scrape in sand or gravel and lining it with quills, bones, fine gravel, and whatever vegetation is present. They lay two dark green-brown mottled eggs, but do not usually rear more than one chick.

Skuas often nest close to penguin colonies; there can be few colonies of pygoscelid or eudyptid penguins which do not have their resident McCormick's or brown skuas. Strongly territorial, the skuas defend an area of penguin nests from attacks of other skuas, and exact a small ground-rent in the form of abandoned eggs, and moribund or dead chicks. Some individuals or pairs, more aggressive than others, trick the penguins into leaving their eggs momentarily, or gang up on healthy chicks and beat down their defences. Where the smaller petrels are breeding, skuas are constantly on the watch for stragglers and sometimes catch adults and chicks as they emerge from their burrows. But much of the food of skuas is krill or small fish, caught honestly at sea or pirated from other birds after an aerial chase.

Both brown and McCormick's skuas have been studied intensively over many years, and the first results of long-term banding studies by William Sladen and others are beginning to emerge. Many young McCormick's skuas return to their natal colony in their second and later years, forming clubs of non-breeding birds. Some begin serious attempts at breeding from their fifth year onward,

Gulls, skuas and terns. The Dominican gull, Antarctica's only species, breeds on the peninsula and all but the remotest maritime and periantarctic islands. McCormick's skua is slightly smaller and greyer than the brown skua, and breeds further south. Kerguelen and Antarctic terns are very similar, differing only in the colour of the outer tail feathers.

BUDD

Brown skuas are predators and scavengers, often breeding among penguins. Here three brown skuas watch closely over a dying macaroni penguin which received its wounds in an encounter with a leopard seal.

but are rarely successful. Even among a cohort of eight-year-old birds in Sladen's study, only half had bred. Breeding success among young birds is extremely low, mostly because they lay later in the season than experienced birds and fall foul of late summer blizzards. Both species of skua return faithfully to nest sites and partners of previous years, though they spend the non-breeding months of May to September wandering widely in tropical and subtropical oceans of both hemispheres.

Antarctic terns are delicate white birds, with grey wings, black cap, and coral-red bill and feet. They breed on Antarctic Peninsula and on almost every island of the Antarctic and Subantarctic zones. Only a single small breeding population has been recorded from the Antarctic continent, at Gaussberg in Wilhelm II Land. Measurements suggest

SPAULL

Greater or wattled sheathbill, a pigeon-like scavenger of penguin colonies and shorelines. Sheathbills breed throughout the maritime Antarctic south to about 65°S. Some individuals winter in South America, others remain near their breeding grounds.

CURTSINGER

Antarctic tern in flight. This species breeds on a wide range of Antarctic and sub-Antarctic islands, feeding inshore, close to its breeding areas. Terns feed by dipping for small fish and other planktonic animals, seldom settling on the water or even wetting their plumage.

that many island populations differ slightly from each other, and several rather doubtful subspecies have been postulated. Antarctic terns feed on small fish and other plankton animals which they catch by dipping, usually on the wing; they seldom settle on the water. During the breeding season they feed just offshore, especially in sheltered water and among pools at low tide.

Antarctic terns nest in small colonies, usually of up to a dozen pairs, forming tiny nest scoops on dry gravel or moraine. Peninsula birds lay two or three eggs, Scotia Arc and more northerly birds only one or two. Both eggs and small chicks match their background very precisely. Adults mob intruders, and are successful at keeping even gulls and skuas from their breeding areas. Juvenile birds have an overall pattern of brown and white barring. After breeding, adults undergo a partial moult, developing a white crown which they continue to wear until their pre-nuptial moult in September or October.

Îles Kerguelen support a second species of tern—the *Kerguelen tern*—which breeds also on Marion Island and Îles Crozet in the Subantarctic zone. Superficially similar to the local form of Antarctic tern, it has slightly darker margins to its tail, and a shorter bill. It nests in small colonies, mostly on turf rather than gravel, and feeds mainly in the intertidal zone or inland, on crustacea, larval and adult insects, and spiders. Kerguelen terns nest in November, slightly earlier than Antarctic terns, and remain over the land throughout winter.

Sheathbills are small, plump white birds, at first glance resembling white pigeons. Their faces, however, bear pink fleshy wattles and a permanently peevish unpigeon-like expression. There are two closely related species, differing in size and distribution, but very similar in ecology and ways of life. The *wattled sheathbill* breeds on Antarctic Peninsula south to about 65°S., throughout the Scotia Arc and on South Georgia. Though often seen about the Falkland Islands and neighbouring coast of South America, especially in winter, they are not known to breed outside the Antarctic zone. The *lesser sheathbill*, slightly smaller and with darker bill, breeds on Heard Island, Îles Kerguelen, and Subantarctic islands of the Indian Ocean.

Sheathbills eat anything; most prominent as scavengers of eggs, dead or dying chicks, and spilled food on penguin colonies, they feed also on intertidal plants and animals, insects, seal faeces and carcases, plankton and other debris washed up on beaches, and any other material which remotely suggests food. Their unwebbed feet scratch busily in the snow, turn over stones and soil, and carry them rapidly out of reach of irate penguins, cormorants and skuas whose eggs they attack. Individuals learn the trick of fluttering up at penguins as they feed their chicks, usually causing food to be spilled—to the advantage of the sheathbill. Though usually seen feeding in small flocks, they nest in solitary pairs, often on high ground overlooking a penguin colony or other major source of food. Two or three eggs, white with grey-brown flecks, are laid in December and January; chicks join the feeding flocks in March or April.

The *South Georgia pintail* is closely akin to the brown pintail of South America, and probably represents a fairly recent natural colonization.

Antarctic land birds. The South Georgia pintail only recently diverged from a South American species, the Kerguelen pintail from a European form. South Georgia pipit is also a recent immigrant from South America. Sheathbills breed only on the maritime and periantarctic islands, wattled sheathbills in the Atlantic sector, lesser sheathbills on the Indian Ocean islands.

LEWIS SMITH

South Georgia pintail. Closely related to a South American pintail, this is one of the very few species of freshwater birds living in the Antarctic region. It feeds in the pools and streams of South Georgia's lowlands.

Common on marshy ground, especially among the broad coastal tussock flats of the warmer, northern side of the island, it hides its nests well at the foot of tussock clumps and feeds busily among the soft, waterlogged vegetation. It lays four to five eggs in November, but only two or three chicks normally manage to survive. The *Kerguelen pintail,* though superficially similar, is likely to have been derived from the north rather than the west; it is probably a local variant of *Anas acuta,* the northern hemisphere pintail which breeds as far north as 80°N. in Canada. It breeds both on Îles Crozet and Îles Kerguelen, forming large flocks in winter. From September onward the flocks disperse, and pairs form among the marsh ground bordering streams and ponds all over the islands. From three to six eggs are laid in well-hidden nests between late November and early January. Inland the birds feed on worms, insects, and plant life; and on small crustaceans and other animals of the intertidal zone.

South Georgia pipits are the only songbirds native to the Antarctic region. Probably descended from the South American pipit *Anthus correndera,* which breeds in Patagonia and on the Falkland Islands, their ancestors were no doubt vagrants carried to South Georgia by the prevailing westerly winds.

South Georgia pipit, the only songbird native to the Antarctic region. This bird breeds on islands off the coast of South Georgia, feeding on insects and seeds on the main island all the year.

LEWIS SMITH

Reddish brown birds, with buff underparts and the characteristic streaked plumage of the pipit, they live close to the ground and are remarkably difficult to spot among the living and dead stems of tussock and other grasses. They feed on insects, amphipods and other small creatures, for which they forage along the beaches and streams. Springtails trapped in the surface tension of small freshwater pools are a favoured food in spring. South Georgia pipits breed in late November and December, mostly on off-shore islands which are relatively free of rats. Their nests are of woven grass, set among tussock roots; three to four green, speckled eggs form the clutch. They remain on South Georgia all the year.

Vagrants and visitors

Antarctic sea birds which breed only on the peripheral islands (South Georgia, Heard and Îles Kerguelen) do not hesitate to fly far south of their breeding latitudes in summer. Light-mantled sooty albatrosses, white-headed petrels and blue petrels, for instance, are regular visitors to the far south and are often seen even over open water inside the zone of pack ice. Similarly, Antarctic waters are visited by Subantarctic and subtropical breeding species. Mottled petrels, which breed on the Subantarctic islands of New Zealand, and soft-winged petrels of the Atlantic and Indian Ocean Subantarctic islands, often appear at the edges of the pack ice in late summer. Sooty shearwaters, breeding on coastal islands of New Zealand, Australia and southern South America, migrate north into the northern Pacific Ocean between breeding seasons, but young, non-breeding birds spend part of their summer in the far south, feeding along the edge of the pack ice in the Pacific and Indian Ocean sectors.

More remarkable visitors are the Arctic sea birds which regularly visit Antarctica. Most numerous is the Arctic tern, a species closely similar to the resident Antarctic and Kerguelen terns, and almost certainly the stock from which they were derived. After breeding in the north between May and September, Arctic terns sweep southward down the coasts of Europe, Africa and the Americas, appearing in offshore waters and over the pack ice in December and January. In non-breeding plumage, they have the distinctive white forehead, when all local terns are black-capped, and often the long feathers of wings and tail are in moult or only partly grown. In March they begin to change into breeding plumage, just as the local resident terns are acquiring their non-breeding white crowns (page 122), and by April all are heading north again toward their Arctic breeding grounds. Pomarine and parasitic jaegers, skua-like predatory birds of the far north, have occasionally been reported from the maritime Antarctic in summer, showing that at least some of these species also migrate from one end of the world to the other during the Arctic winter.

Casual visitors include upland and white-rumped sandpipers, and white herons—Arctic migrants to South America, which have appeared from time to time in the Scotia Arc region. The South Shetlands have also received ruddy ducks and black-necked swans from southern South America. Îles Kerguelen also record an occasional Arctic migrant (greenshank), and less improbable though equally exotic visitors—a frigate bird (probably a greater frigate from the Indian Ocean) and a broad-billed roller from Madagascar.

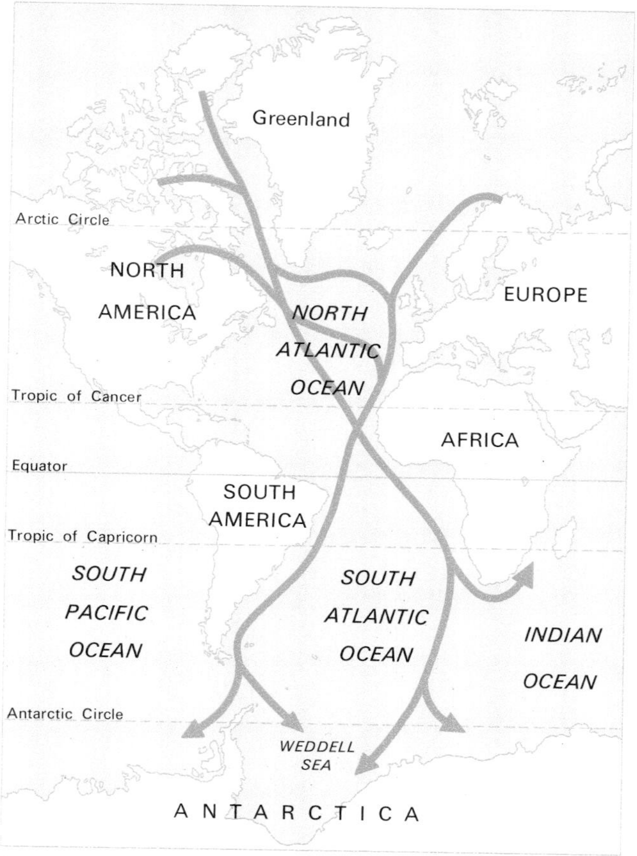

Migration route of Arctic tern. Arctic terns breed in high latitudes of the northern hemisphere, each year between May and September. Moulting into non-breeding plumage, they fly southward in small bands down the coasts of Europe, Africa and the Americas, to spread east and west into Antarctic and Subantarctic waters. Many feed among the pack ice. In March and April they begin a prenuptial moult and return to their northern breeding grounds.

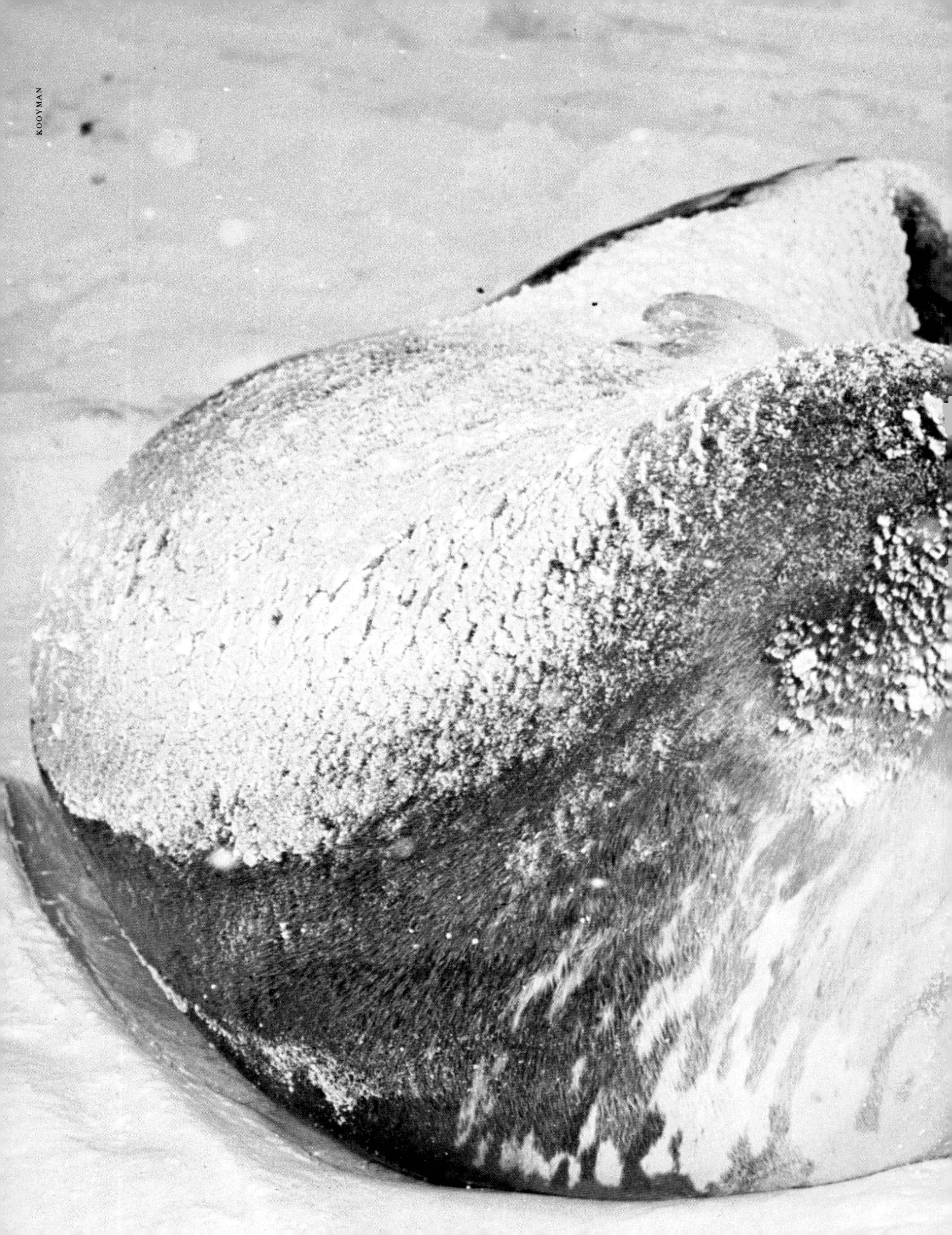

KOOYMAN

The Mammals of Antarctica

Though continental Antarctica and the older Antarctic islands must once have possessed a lively fauna of land mammals, all disappeared without trace before the advance of the ice cap in late Pliocene and Pleistocene times. The newer Antarctic islands—Heard, Bouvetøya, the South Sandwich groups and other volcanic upstarts, have had no links with other land masses and no chance to acquire land mammals of their own. Excepting those which man himself has introduced, all Antarctic mammals are therefore marine mammals. They include twelve species of whales and six species of seals, some plentiful—or formerly plentiful—and the subject of many years' intensive research, others rare and almost unknown to man.

Previous page: During blizzards on the sea ice, Weddell mothers make a windbreak for their pups. Their own insulation of blubber is so efficient that in spite of a high body temperature, powdered snow remains unmelted on their surface.

Blue whales, for long the major prey of pelagic hunters, were already becoming scarce before World War II after the intensive hunting of the 1930s. Fin whales took their place, but they too began to decline. Now the pressure has fallen upon sei whales, previously judged too small to be worth hunting.

Whales

The whales of the world fall into two groups, both well represented in Antarctic waters. Whalebone or baleen whales are generally large animals, ranging in length from 10 to 30 metres. They live pelagically, seldom diving far below the surface, and feed on plankton. Lacking teeth, they grow instead a sieve of springy whalebone plates, which hang from the roof of the mouth like leaves in a book. The

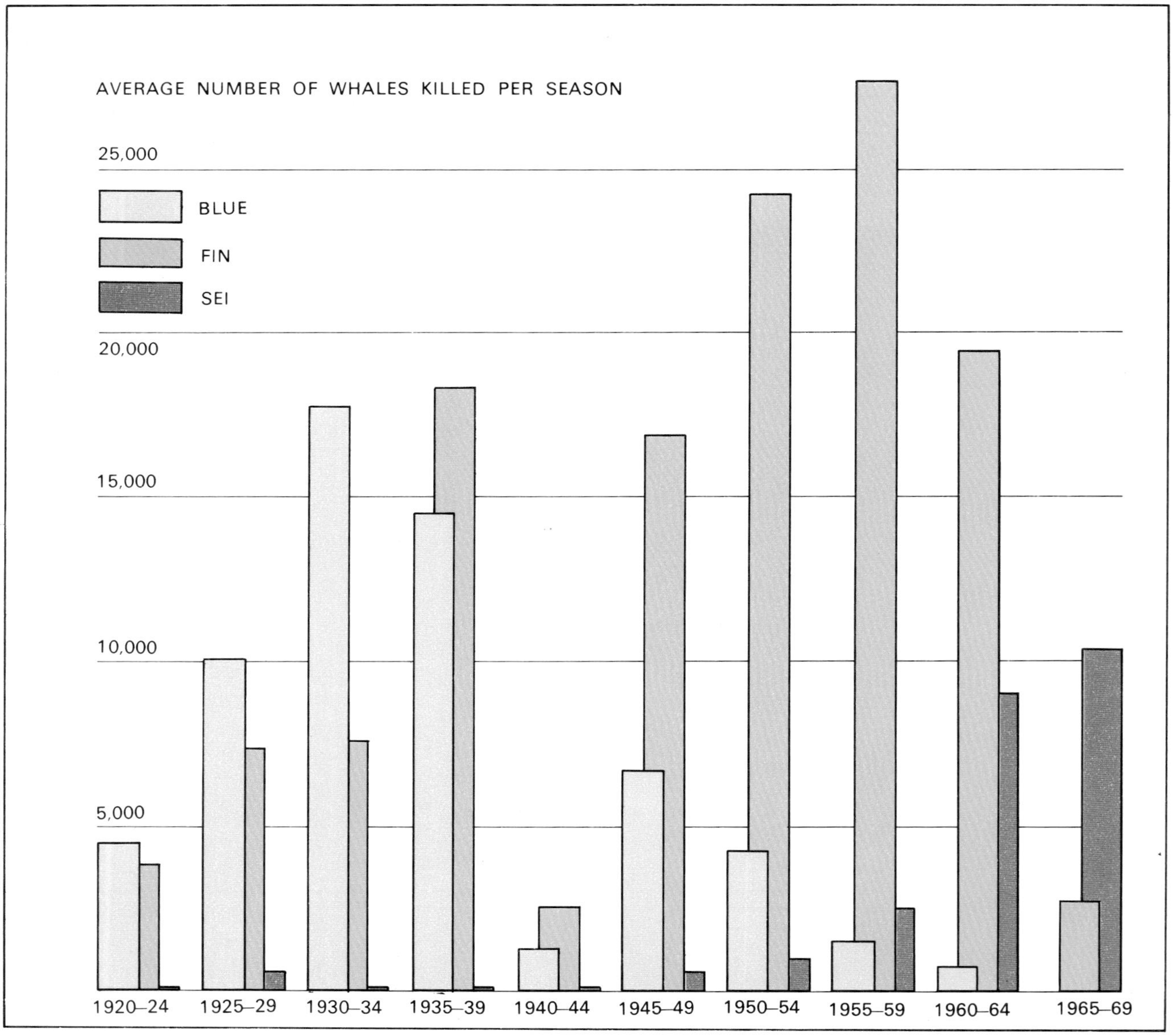

FLOOKES

frayed edges form a filter with the texture of coarse coconut matting. Swimming steadily through plankton shoals, the whales take in huge concentrations of plankton and sea water. The water is pressed through the matting, leaving the plankton behind to be swept down the throat by the huge tongue. Toothed whales have orthodox teeth, ivory crowned though never arranged in the orthodox mammalian pattern of incisors, canines and molars. Sperm whales, largest of the toothed whales, have up to twenty-six peg-like teeth on either side of a long narrow lower jaw, but none in the upper jaw. Dolphins have as many as forty per side in upper and lower jaw, killer whales and porpoises twenty per side, and bottlenosed whales one or two per side, in the lower jaw only. They feed mostly on fish and squid, but killer whales also attack porpoises, seals, and even large whales. Toothed whales often dive deep in search of prey, and use a repertoire of sound signals in hunting and communication with each other. They are highly social; killer whales hunt in what seem to be enhanced family groups; dolphins and porpoises often form huge schools of several hundred.

Whales as a group are air-breathing mammals, extraordinarily well adapted for life at sea, and perfectly at home in the coldest polar water. Originally four-legged, they have lost all trace of hind limbs, and use their fore limbs as paddles and hydroplanes. The flat tail flukes form a rotary propeller, which drives the animals forward comfortably at the speed of a slow cargo boat. Over short distances they can travel much faster, but they soon tire and become puffed when chased. The skin is rubbery, hairless except for a few sensory bristles on the face, and underlain with blubber—thick, fibrous fatty tissue, up to 15 centimetres in most species but four times as thick in right whales (page 134)—which is both a food store and an insulating layer. Whales maintain body temperature in cold water partly by the efficiency of their insulation, and partly by sheer size; they are large animals with tremendous capacity for producing energy, and only a small surface through which to lose heat. Skin and blubber are well supplied with blood. When the animal is still and generating little surplus energy, the blood vessels of the surface layers contract and both skin and blubber become effective insulation. When the whale is active and producing heat rapidly from its muscles, skin and blubber flush with blood and help to release surplus heat into the water.

Blubber contains a clear, straw-coloured oil which, until mineral oils replaced it, was much in demand as a fuel and lubricant. Whalebone, too, had a thousand uses before cheap spring steel and plastics were invented, and whales have been hunted for their whalebone and oil for over three centuries. Commercial whaling began in Europe and spread to North America during the eighteenth century. First of the large whales to be harpooned were slow-moving right and humpback whales, which were sluggish enough to be tackled from small open boats, did not sink when killed, and carried good measure of whalebone and oil. Sperm whales, largest of the Odontocetes, have no whalebone but are well covered with blubber and carry a liquid wax—spermaceti—used in candle-making, lubrication, pharmacy and cosmetics. During the eighteenth and nineteenth centuries a vast whaling industry grew up about these species, based in the northern hemisphere but spreading across the southern Pacific, Atlantic and Indian Oceans.

Rorquals—the streamlined, fast-moving blue, fin, sei and minke whales of both hemispheres (see below)—were immune to attack during this early phase of whaling. Too lively and fast to be taken from open boats, they tended to sink when killed, taking their captors with them, and were altogether too much for the primitive whalers to handle. However, the explosive harpoon, developed in 1864 and mounted on swift, steam-driven catchers, boosted the industry and began the destruction of the rorqual. Within a century blue whales, largest and most sought-after of the rorquals, were almost completely destroyed at both ends of the earth, and fin whales were rapidly heading toward elimination. Antarctic whaling, beginning in 1904, virtually collapsed in the mid-1960s when the larger rorquals had become extremely scarce and hunting no longer paid.

A fraction of the profits from Antarctic whaling was used to finance whale research in southern waters during the 1920s and 1930s, and again after World War II. In consequence, the biology of the southern rorquals is well known. *Blue whales,* the world's largest animals, reach lengths of 30 metres and weights exceeding 100 tonnes; the mean length of Antarctic blue whales is about 24 metres, the mean weight about 84 tonnes. Slate grey all over, they often carry a film of yellow diatoms which gives them a bluish appearance in the water. *Fin* or *finback whales* are a size smaller, growing to 25 metres but averaging 20 metres and weighing about 50 tonnes. Pale grey above with white throat and belly, fin whales glide like pale ghosts through the

BALEEN WHALES

Blue whale: mean length 24 metres, mean weight 84 tonnes.

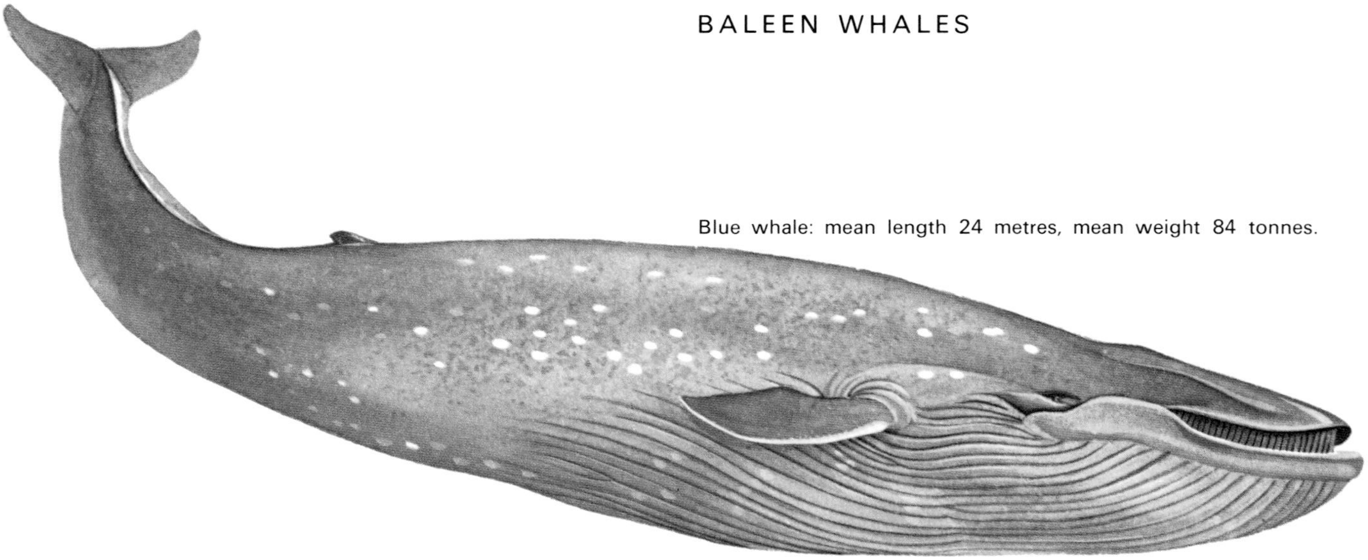

Fin whale: mean length 20 metres, mean weight 50 tonnes.

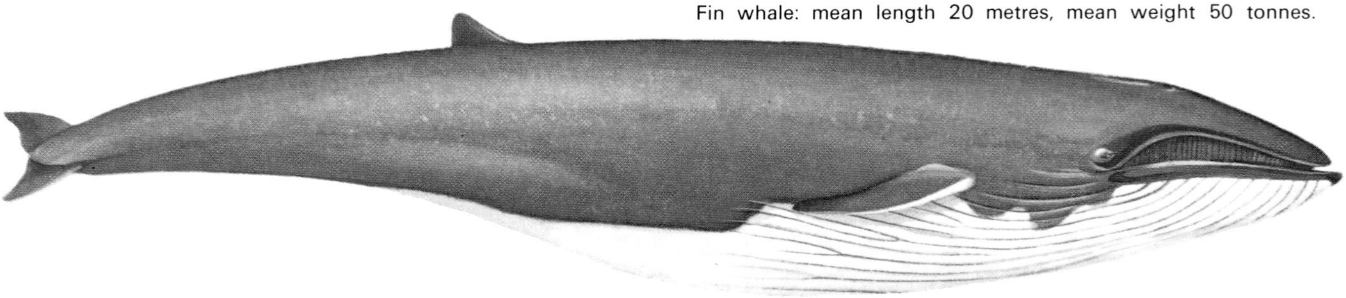

Sei whale: mean length 16 metres, mean weight 13 tonnes.

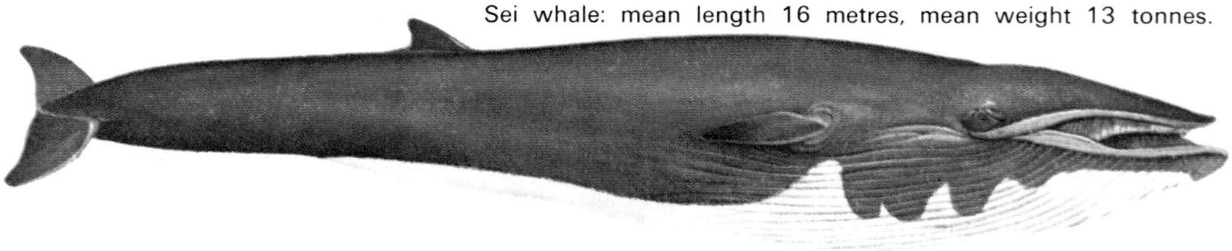

Southern right whale: mean length 20 metres, weight unknown.

COOK

Humpback whale: mean length 18 metres, mean weight 33 tonnes.

Minke whale: mean length 9 metres, mean weight 7 tonnes.

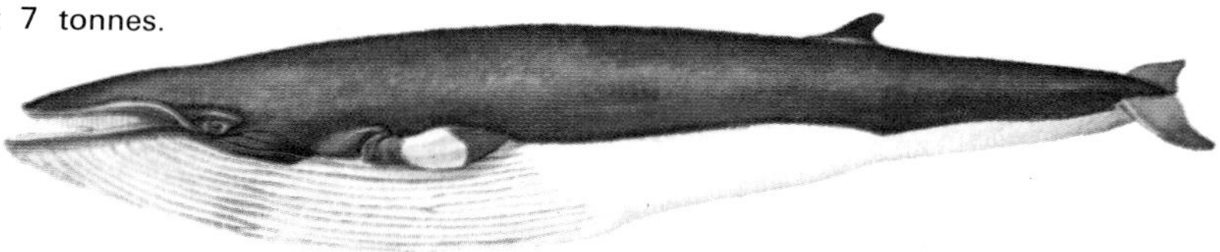

The whalebone or baleen whales (left and above) are large and tend to stay close to the surface of the water. Their "whalebone" is sheets of springy, fibrous material with frayed edges, through which planktonic animals are filtered from the sea. The toothed whales are generally smaller—only sperm whales match the baleen whales in size—and take bigger prey, including fish, squid, and even seals. Killer whales especially have a reputation for ferocious hunting. Sperm and bottlenosed whales, especially, dive deep for their food, but all whales must come up to the surface to breathe.

TOOTHED WHALES

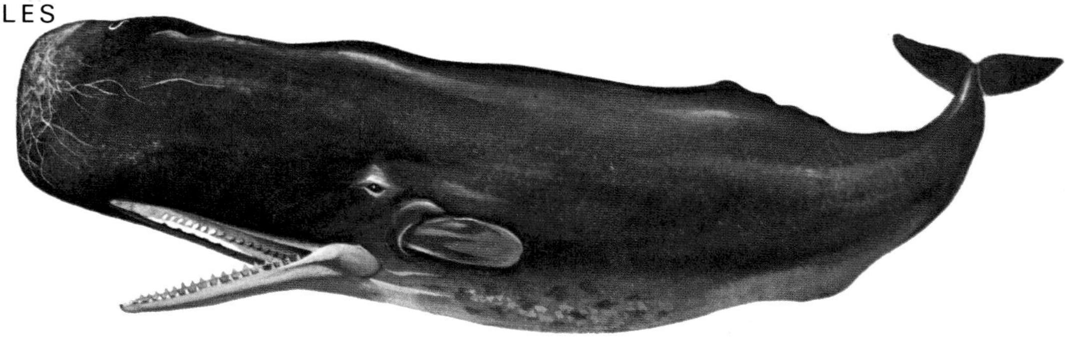

Sperm whale (males only in Antarctic waters): mean length 16 metres, mean weight 35 tonnes.

Bottlenosed whale: mean length approximately 10 metres.

Blackfish: mean length of males 4 metres, females slightly smaller.

Killer whale: males average 9 to 10 metres, females $4\frac{1}{2}$ to 6 metres.

Dusky dolphin: mean length $2\frac{1}{2}$ metres, weight unknown.

Cruciger dolphin: mean length 2 metres, weight unknown.

Southern bottlenosed whale: mean length approximately 7 metres.

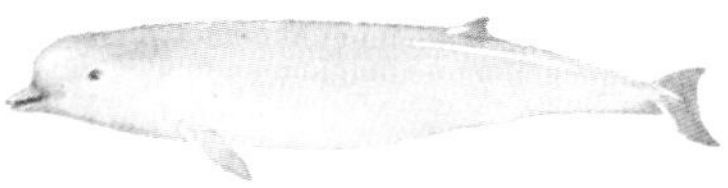

Spectacled porpoise: mean length 2 metres, weight unknown.

dark Antarctic water. *Sei whales*, still smaller, seldom exceed 15 metres and average 22 tonnes. They are dark grey or black, with a narrow white band underneath. *Minke whales*, smallest of the rorquals, reach lengths of 10 metres and only half the weight of seis. Similar in colour and pattern to the rest of the family, they have a broad white stripe across their flippers, and are white underneath from chin to tail. Rorquals of the northern hemisphere are similar to these, but slightly smaller.

All the rorquals have short flippers, a mouth extending back one-quarter to one-fifth the length of the body, and a throat and abdomen pleated with longitudinal grooves. Nobody knows what the pleats are for. They may help to increase the surface area of the skin and allow more heat to escape when the whale is swimming fast. They may allow the whale to distend its throat while gulping food or breathing deeply, or they may add to its fore-and-aft streamlining and smooth its passage through the water. Rorquals feed on zooplankton, almost entirely on the large, reddish euphausiid crustaceans collectively called *krill* (page 134). The different species of rorqual may select shoals of different age groups and species composition to feed on. Blue whales are solitary creatures, sometimes seen in pairs but seldom in groups. Fin, sei and minke whales are more gregarious, generally swimming in large or small schools.

Southern rorquals migrate south into Antarctic waters each spring, arriving thin and hungry after several months of poor feeding in subtropical waters. Females usually have small calves with them, born in warmer northern waters and still feeding on their mothers' rich milk. Blue whales head south first, closely followed by fin whales and the smaller species. Blue, fin and minke whales feed far south along the edge of the pack ice; minkes often enter the pack and feed in its leads and openings—occasionally finding themselves in trouble when the pack freezes in late autumn (page 40). Sei whales seldom travel as far south as the other three species. Rorquals feed voraciously and fatten throughout the summer. Whalers like to catch them late in the season, when they have had time to lay on good stocks of oil from the rich plankton.

Humpback and *right whales* are almost as long as sei whales, but fatter and heavier, with more blubber and a greater yield of oil. Humpbacks are ungainly, lumpy whales with long, knobbly flippers and serrated margins to flippers and tail. Dark above and white below, they have a pleated throat (with fewer and broader pleats than rorquals) and

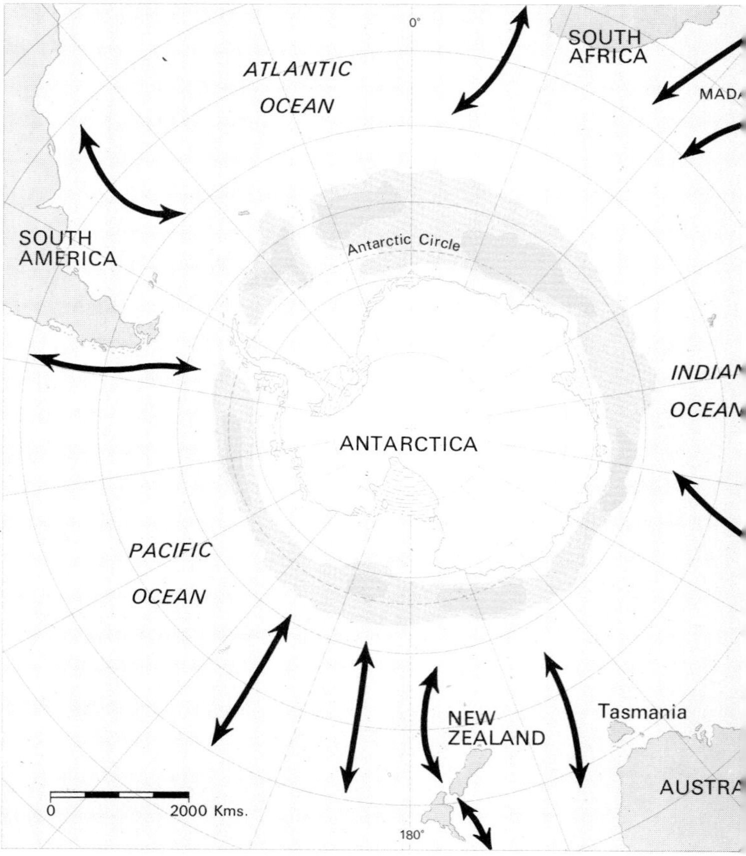

Whale hunting areas. Whales gather where their food is most plentiful. They are especially numerous in areas where climatic or other disturbances mix water masses, e.g. in the zone of westerly winds, in the massive eddies which form "downstream" from South Georgia and the South Sandwich islands, and near the Kerguelen Ridge.

Right: Fin (foreground) and sei whales drawn up on afterplan of a whale factory ship ready for flensing. Short plates of whalebone or baleen fill the upper jaw of each.

Minke whale bursting a hole in new ice to "blow". The jet of warm gases blown from the lungs can be seen behind the head.

KOOYMAN

SWITHINBANK

Krill (*Euphausia superba*) from the stomach of a baleen whale. These small shrimp-like animals, 6 to 9 centimetres long, are gulped in shoals by the whale and filtered by the mat-like baleen

MERRETT

Head of a sperm whale showing the row of peg-like teeth in the narrow lower jaw and the enormous "casket" of spermaceti forming the front of the head

a short, broad head. They swim slowly with a leisurely roll, some slowly enough to grow barnacles. However, they make annual migrations between tropical and polar or subpolar waters. In the southern hemisphere their migrations take them inshore along the coasts of Australia, New Zealand, South America and South Africa, where they are especially vulnerable to attack from coastal whaling stations. Humpbacks were the first prey of whalers working from South Georgia and Îles Kerguelen, and have also been hunted extensively in the open ocean.

Southern right whales, even fatter than humpbacks, have an enormous head—nearly one-third the length of the body—with a deep unpleated lower jaw and baleen plates up to 4½ metres long. Only marginally an Antarctic species, they were once widespread in waters from 15° to almost 60°S. Slow-moving, and buoyant because of their massive blubber, they were the favourite quarry of the open-boat whalers. Their habit of calving in sheltered inshore waters, where both mother and calf could easily be taken, also recommended them to the hunters, and southern right whales were rare even before mechanized whaling began. Very few remain in the southern oceans today.

Sperm whales, largest of the Odontocete whales, visit Antarctic waters regularly each summer. Only males occur in the far south. Up to 18 metres long, with mean weight of 35 tonnes, they are remarkable for their enormous "casket" of spermaceti, which fills the bulging forehead. We do not know what it is for. Its buoyancy helps the head to rise in the water, but there may well be other functions, too. Females, only half the length of the males, stay in temperate or warm oceans throughout the year. Sperm whales feed almost entirely on squid, which they probably catch in very deep water. Often they bear the scars of squid suckers on jaws and face. They dive to 400 metres or more, and can stay below the surface for over an hour. Sperm whales are hunted throughout the world, but the large males of polar waters are especially sought after. Of 20,000 taken each year in all the oceans, one-quarter are killed in the Southern Ocean.

Killer whales, like minke whales, are too small to be chased by large-scale commercial hunters. Fully grown males are only 10 metres long, and females seldom reach more than 6 metres. Killers are strikingly marked in black and white, with a distinctive white patch behind each eye and often a grey patch behind the dorsal fin. Large males have a tall dorsal fin almost 2 metres tall, which sticks out of the water like a triangular sail when they surface to blow. They move about in packs, sometimes of thirty to forty animals, and in the folklaw of the sea have earned a reputation for cunning and ferocity. Killer whales feed on porpoises, seals, and even larger whales; they are reported to emerge among ice floes, rolling the seals off into the water. Plentiful in the Antarctic, they sometimes make a nuisance of themselves by tearing chunks off recently-killed rorquals, and are shot for their pains by the whalers. Individual killer whales in captivity seem to be charming beasts, who jump through hoops, stand on their head and allow their teeth to be brushed without comment.

Our small knowledge of the lesser toothed whales has been summarized by Dr. Francis Fraser. The *spectacled porpoise*, which grows to a length of 2 metres, is known from only a few specimens collected in South America, the Falkland Islands and South Georgia. Intensely black and white, its white upper lip and the white ring about the eye are distinctive. Up to forty-two spade-shaped teeth in upper and lower jaw suggest a diet of fish. The *cruciger dolphin*, first noted and named in 1824, was never fully examined until a specimen—only the fourth on record—was caught by whalers and sent to the British Museum (Natural History) in London in 1960. This is a small, energetic dolphin some 2 metres long, with delicate grey, black and white tracery on its flanks and a dark belly-band, which seems to keep company with groups of fin whales at the edge of the pack ice, and extends north into cool temperate seas about Cape Horn. *Dusky dolphins* and *blackfish* are cosmopolitan species; skulls have been collected at Îles Kerguelen, but there are only occasional sight records—always suspect in turbulent waters—to suggest that they penetrate further south. Other dolphins, e.g. the "hourglass dolphin" described by Edward Wilson from the deck of a moving ship, have been recorded in Antarctic waters and may well exist, but their identity is suspect until specimens have been properly examined and compared with known species.

Bottlenosed whales, with their curiously elongate snout, rounded forehead and single or double pairs of triangular teeth, are more certainly known. The *southern bottlenose*, which may reach a length of 7½ metres, has been recorded at the South Orkney Islands, South Georgia, and in the open ocean in 61°S. Its curious dentition suggests a diet restricted almost entirely to squid. *Berardius arnouxi*, a slightly larger bottlenosed whale without a common name, has been recorded at the South Shetland Islands and off the northeastern tip of Antarctic Peninsula. Julian Taylor, biologist at a British base, reports that in April 1955 a single specimen of this rare whale was trapped by newly-formed sea ice in Prince Gustav Channel, between Ross Island and the tip of the Peninsula. With it were minke and killer whales, who lived in harmony, and whose combined activities kept a few square metres of water open for breathing throughout the winter. Visitors from passing sledge parties stopped off to pat the whales, which stuck their heads up through the hole and seemed to remain healthy in spite of the confinement. In August several of them, including the bottlenose, were seen jumping in their pool, leaping over two metres into the air. Taylor reports that, shortly before the ice broke out and relieved the captives, the bottlenose was fired on by Argentinian soldiers from a nearby base and presumably killed.

Killer whale surfacing to breathe. The distinctive white patch behind the eye shows clearly and the open blow hole or nostril shows the dorsal surface immediately above it.

KOOYMAN

LEWIS SMITH

Weddell seal mother with twin pups, rare in this species. The hind flippers of phocid seals cannot be turned forward like those of fur seals and phocids do not raise their body from the ground when moving.

Young fur seal with tagged fore-flipper. The Monel metal tag, clipped to loose skin behind the flipper, should remain in place for several years. It carries a number which identifies the animal.

TICKELL

Seals

The seals of the world fall into three families, of which two—the Otariidae (fur seals and sea-lions) and Phocidae (true seals)—are represented in Antarctic waters. The third family, Odobenidae (walruses) live only in the Arctic. Like whales, seals evolved from four-footed land mammals, the otariids possibly from bear-like ancestors, the phocids from otter-like forms. Unlike whales, seals are not exclusively aquatic. They feed and spend much of their time in water, where they swim easily though not with the effortless grace of whales. Retaining all four limbs, they have remained to varying degrees mobile on land. Otariids walk with the body off the ground, limbs pulled well under, and flippers sticking out awkwardly to the side like oversize comic boots. Surprisingly efficient, they run, scramble and climb over rocks with agility; an irate bull fur seal can outrun a man on rough scree. Phocid seals are more slug-like, with shorter fore-flippers and trailing hind limbs. Both on land and at sea they move by undulating the body muscles. At sea the combined tail and hind limbs form a broad-bladed propellor. On land they slither, roll and undulate like very fat but agile

Elephant seal. Largest of the Antarctic seals, male elephant seals measure 6 metres or more and weigh up to 4 tonnes. Females seldom exceed $3\frac{1}{2}$ metres.

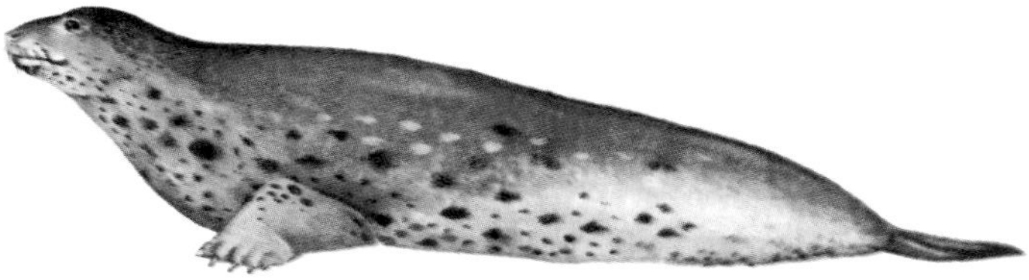

Leopard seals are lithe, slender hunters, found mainly on the pack ice but also on the mainland coasts and widespread islands of Antarctica. They measure up to $3\frac{1}{2}$ metres long; males and females are of similar size.

Weddell seals are fat, sluggish seals of the continental shore and southern Scotia Arc. They feed mainly on fish, which they catch by diving deep through holes in the fast ice. Large animals of either sex measure 3 metres and weigh up to 400 kilograms.

caterpillars. Leopard seals slither over snow fast enough to chase penguins when the mood takes them, and wise explorers approach them with caution. Weddell seals are slowest and most somnolent. Unused to enemies on land, they lie quietly when man approaches and, with only mild protest, allow him to paint numbers on them for identification. Elephant seal bulls up to seven metres long charge each other in battle like huge overwrought frankfurters, pausing for long rests between rounds.

Seals emerge from the water to produce their young. Some spend long spells ashore in mating, tending the pups, moulting, and loafing in the sun. For all these purposes sea ice is an adequate alternative to land, and at least three Antarctic species live their whole lives at sea or on ice without coming ashore. All seals feed at sea, hunting by sight and possibly by sonar. Their large eyes have an almost spherical lens, which helps their underwater vision, and in some species a reflecting layer behind the retina helps to improve vision in poor light. Weddell seals, almost silent on land, become very vocal in the water. Their trills and squeals may help them to echo-locate food in the dim world under the fast ice, and find their way back to the

Crabeater seals are restricted almost entirely to the pack ice. They feed mainly on krill (animal plankton) which they sift from the sea through many lobed teeth. The sexes are of similar size, large animals measuring up to 3 metres.

Ross seals are comparatively rare animals of the pack ice, and little is known of them. They are slightly smaller than other phocid seals of the Antarctic, measuring on average 2 to 3 metres. They probably feed mainly on fish and squid.

Fur seals breed only as far south as the southern Scotia Arc islands, and are almost unknown among pack ice or coastal ice of the continent. Their luxuriant fur would perhaps be disadvantageous in extreme cold. Males measure up to 1·8 metres; females up to 1·5 metres.

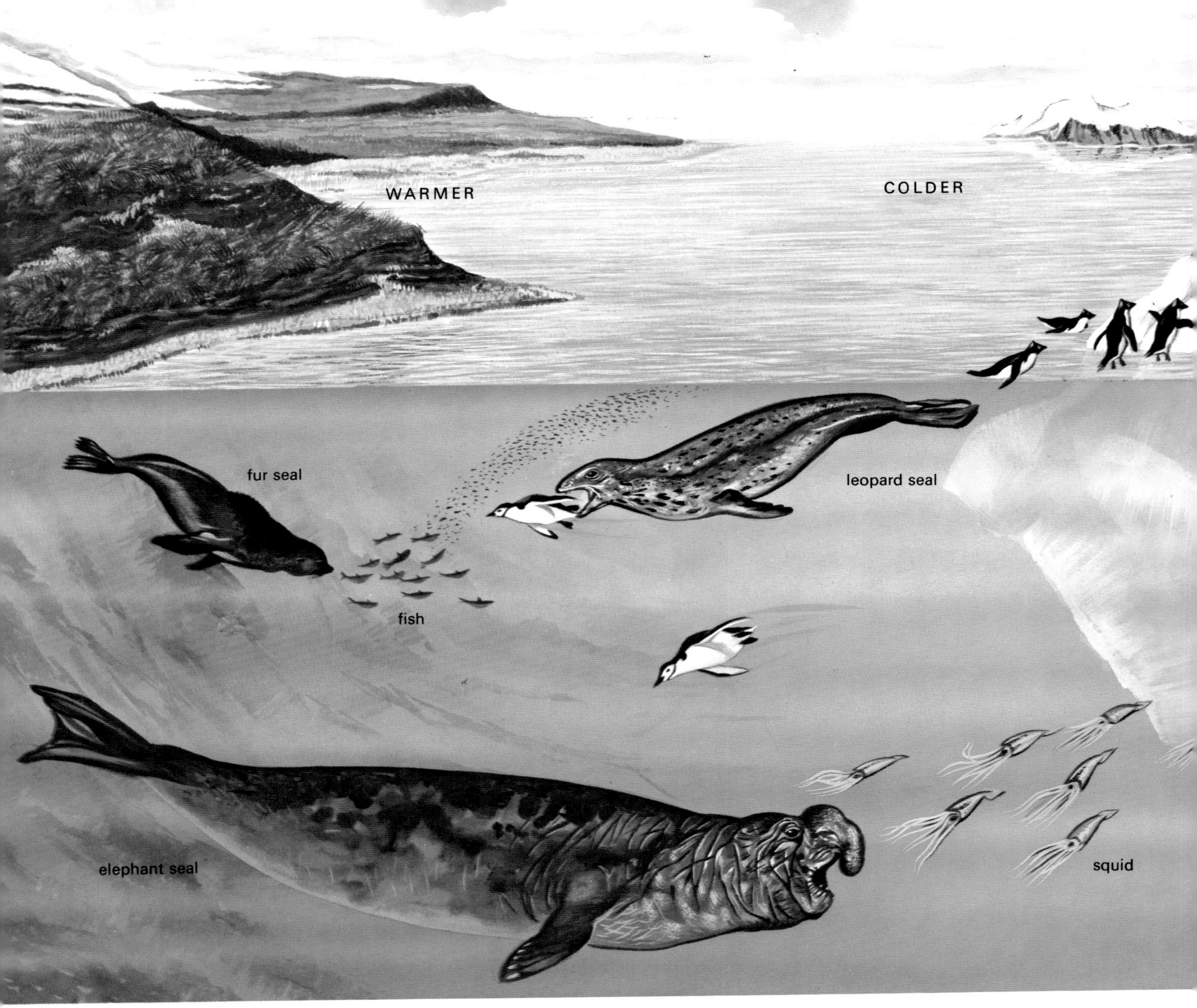

The six species of Antarctic seals divide the environment up between them; no two species feed entirely in competition with each other, though several species may from time to time feed side by side on a locally abundant prey. Weddell seals, a high Antarctic species of the continental and maritime zones, feed mainly on fish caught at depths under the ice. Crabeater, Ross and leopard seals, more typically found on pack ice, feed respectively on krill, fish or squid (probably), and a diversity of surface-living creatures including fish, krill and penguins. Elephant seals live mainly in the maritime and periantarctic zones, and feed mostly on fish and squid. Fur seals, who live furthest north in the warmer maritime and periantarctic zones, feed on krill and fish, no doubt taking squid when they are plentiful.

These skulls show a number of interesting adaptations which fit the different species for their ways of life. The massive jaws and peg teeth of the elephant seal provide weight in fighting: the deep rooted canines may be especially important, and grow persistently throughout life. Teeth of leopard seals are sharp and well fitted for cutting flesh. Some of the molars, like those of crabeater seals, have prominent cusps which may be used in filtering. Weddell seals have small, peg-like teeth which are probably

adept at holding fish. The protruding incisors also help in gnawing the edges of ice holes in winter. Ross seals have narrow, needle-like incisors, perhaps right for holding slippery squid. Fur seals have small, general-purpose teeth with no obvious specialization. Note the enormous eye sockets of the Ross seal, accommodating very large eyes.

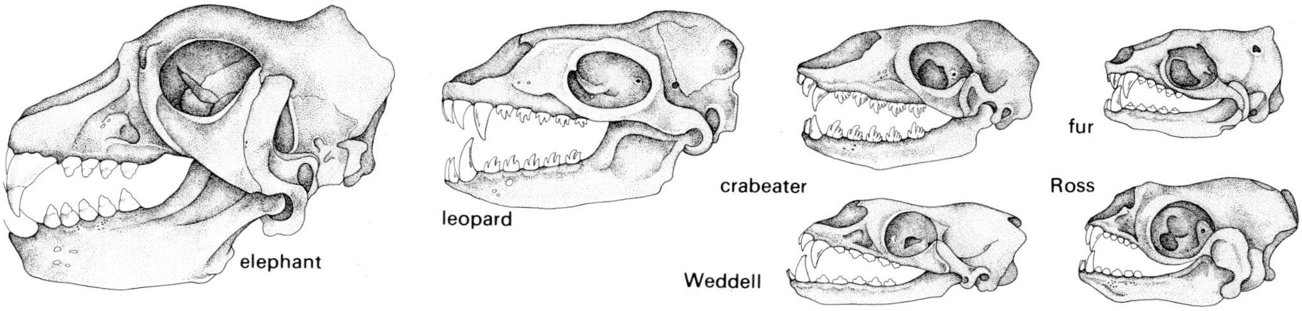

TICKELL

Newborn elephant seal pup still in its amnion or birth membrane. The dense, black, woolly fur keeps the pups warm for the first month of life and is later moulted for a shorter, silver-grey coat.

breathing holes. Some seals seem especially well adapted for deep diving and long spells of work without surfacing. Elephant seals, which may well dive deepest, can work for twenty minutes or more under the water. Weddells hunt at depths down to 600 metres, and stay down for over an hour.

Fur seals Of the seven closely related species of southern fur seal (genus *Arctocephalus*) only one penetrates south into polar water. It is a southern subspecies of *Arctocephalus tropicalis*, a species which breeds on either side of the Antarctic Convergence in the Indian Ocean, and appears to have spread fairly recently into the maritime region of the Antarctic. The southern subspecies, first described from Îles Kerguelen and called the *Kerguelen fur seal*, breeds also on Heard and Macdonald Islands, Bouvetøya and throughout the Scotia Arc. At all its breeding stations numbers have increased substantially in the past few decades, and the total population may now have reached 50,000. Thus many islands which formerly held large fur seal populations, but were cleared of seals by the hunters of the eighteenth and nineteenth centuries, resound once again to the yaps, roars and squeals of fur seal breeding colonies.

Kerguelen fur seals are now widely distributed throughout the Scotia Arc. Though increasing in numbers, stocks are still far below those of the early nineteenth century when hundreds of thousands were killed for their skins.

LAWS

On South Georgia, where they have most systematically been studied, fur seals assemble at the breeding beaches from late September onward. Large bulls, up to 1·8 metres long and weighing over 150 kilograms, take up and defend small territories against each other. In October the females return from sea. Measuring only 1·5 metres, and less than half the weight of the bulls, they gather on the territories in harems of four or five. The bulls defend them energetically against the attentions of young unattached males and their neighbours. Pups are born within the harems in late November and early December. Like small, woolly dogs they yap and growl at each other, rolling and fighting in the mud of the colonies. Only 0·45 metres long at birth, they grow steadily through the summer, moulting in April into an elegant silver-grey fur. Weaned at the same time, they leave for the sea. The mothers are impregnated about one week after the birth of their pups, but the developing blastocyst does not settle and embed in the wall of the uterus until March or April. This ensures that the females, with a gestation period of seven months, give birth at the right time of year. Adults moult between February and April, and spend much of the winter at sea.

Weddell seals, like other aquatic mammals, must return to the surface to breathe. Those living in the far south use their teeth to keep holes open in the ice. They feed on fish and squid which they probably catch at depths down to more than 100 metres.

TICKELL

EVERSON

Fur seals are covered with a coarse outer layer of guard hairs which easily shake free of moisture and snow. Beneath is a velvety layer of very fine texture, the fur prized by furriers. Guard hairs are removed when the pelt is processed.

Right: Bull elephant seal threatening with inflated proboscis. Large bulls measure up to 6 metres long and probably weigh 3 to 4 tonnes.

A fur seal emerging from the sea carries a glaze of water which will shortly be shaken off. Fur seals live only in the maritime and warmer parts of the Antarctic region, seldom encountering very low temperatures.

Elephant seals Largest of the phocids, elephant seals breed in southern South America and on practically every Subantarctic island which has room for them. In the Antarctic they are widespread, breeding on Heard, Macdonald, Îles Kerguelen and South Georgia, and throughout the Scotia Arc. Like fur seals, elephant seals are still recovering from the slaughter of the nineteenth and early twentieth centuries; well established at all their breeding islands, they seem, if anything, to be spreading and increasing in numbers. World population is estimated at about 750,000.

The coarse hair of elephant seals, only 1 to 2 centimetres long, grows from a remarkably tough hide. Underlying blubber is 5 to 10 centimetres thick in cows, up to 15 centimetres in large bulls, and may account for one-third of the total weight of the animal. Large, fully grown adult males measure 6 metres and probably weigh 3 to 4 tonnes in full fat. Females seldom exceed lengths of $3\frac{1}{2}$ metres. Elephant seals spend much of their winter at sea, returning to the breeding grounds—usually sheltered sand or shingle beaches—from early September onward. Mature bulls take up beach territories, and the returning cows gather about them in huge, indeterminate harems. Only mature bulls of seven years or more are large enough to hold even a corner of a harem. Most breeding bulls are twelve years old or more, and at least 5 to 6 metres long. The bulls roar and fight continuously while the harems are forming. Equipped with inflatable nostrils which hang over the mouth like a trunk and form a resonating chamber, they produce a sound which deafens at close quarters

CONROY

Elephant seal pups, one to two weeks old. Born in October or early November, pups feed only on milk, trebling their birth-weight of 40 to 50 kilograms, in three weeks.

Right: After breeding, elephant seal cows fatten and gather in groups on the beaches to moult. The fur strips off in patches. At a time when the new fur is growing and the skin is well supplied with blood, grouping probably helps to reduce heat losses.

and, accompanied by violent displays of aggression, keeps rivals at a distance. Fights occur frequently; weight rather than skill seems to count, and the larger beast usually remains holding the territory. The males fast throughout the breeding season, living on their prodigious reserves of fat.

The first pups are born in October, usually within a week of the cow's coming ashore. Later arrivals pup as late as early or mid-November. Weighing 40 to 50 kilograms at birth, the pups are fed on a viscous milk rich enough to treble their birth weight in three weeks. The mother remains ashore throughout this period, losing some 300 kilograms of fat, protein and liquid from her stored reserves. Oestrus and mating occur as the pup is weaned, and the mother leaves for the sea soon afterwards. The pups live on their fat for three to four weeks, completing their first moult, then depart for the sea themselves. Adults return in late summer to moult, lying nose to tail for three to four weeks on the beaches and in malodorous wallows among the tussock grass. Skin and hair peel off in huge flakes, and the animals scratch, grumble and quarrel among themselves in garrulous discontent.

Elephant seals of South Georgia were exploited for many years by one of the whaling firms of South Georgia, under licence from the Government of the Falkland Islands. Large males, of which there is a surplus in the population, were taken each year in September and October, up to a total of 6,000. Only three-fourths of the island's beaches were open in any season, only bulls exceeding 3·5 metres long could be killed, and a proportion of mature bulls was left on every beach. This controlled exploitation appears to have worked well; though possible changes have been noted in breeding behaviour and other aspects of the species' life, the population was maintaining itself and apparently flourishing when the industry ceased in 1964.

Weddell and crabeater seals *Weddell seals* are the true seals of continental Antarctica, breeding and hauling out mainly on fast ice close to the mainland coast. Further north they gather wherever reliable fast ice forms in winter, notably along the west flank of Antarctic Peninsula and on the South Orkney Islands. The most northerly breeding colony is at Larsen Harbour, a narrow, land-locked inlet of South Georgia, where Weddell seals seem perfectly at home on flat beaches. The world population is difficult to assess because so many breeding areas are seldom or never visited; it probably lies between half and one million.

Weddell seals are gentle creatures, measuring up to 3 metres long and weighing up to 400 kilograms. Dark grey on top and pale underneath, their coats are flecked and streaked with silver, and often trimmed delicately with hoar-frost on misty mornings. In summer they sleep long hours on the sea ice, usually grouped about a breathing hole which gives them access to the sea. During cold spells in spring and autumn they keep the hole open

by rasping its edge with their teeth. Many older animals have worn and broken their incisors and canines almost to the gums, and suffer badly from dental abcesses deep enough to erode the bones of their jaws. In winter Weddell seals spend much of their time under the ice, avoiding the coldest weather and the strong winds. They are especially numerous where pressure forces the ice into hummocks, providing cracks and sheltered caverns for them under the surface sheets.

In the South Orkney Islands they begin to gather on the breeding grounds—usually stable stretches of inshore fast ice—in July and August, and the pups are born in September. Further south, assembly and pupping occur one to two months later, and may be delayed by unseasonable weather in October and November. Covered at first with long woolly hair, the pups weigh 30 to 35 kilograms and are about half as long as their parents. The mothers, normally placid, defend their offspring against neighbours, and will even attack man if he comes too close. The pups enter the water during their second week, but are suckled and tended by their mothers for a further month or more. Parent and young swim together; often the mother cuts a ramp in the ice for the pup to emerge, and helps it out of the water with nudges and calls. Males take no obvious part in rearing the young, though they are sometimes seen lying on the ice among the groups of mothers, and occasionally fight both at the surface and under water. Their presence may be necessary to keep interlopers away and reduce competition for space and food about the breathing holes. Mating occurs in the water, about a month after the pups are born, and implantation is delayed by a further month.

Weddell seals have seldom been hunted by commercial sealers, except at their more northerly breeding grounds where colonies were occasionally rounded up for the try-pots. Many hundreds have been killed for dog food by expeditions which use sledge-dogs in the field. Ian Stirling, who studied the effects of harvesting on a local population of Weddell seals at Scott Base, McMurdo Sound, found that the removal of a high proportion of animals of breeding age between 1956 and 1960 (when seals amounting to about a quarter of the breeding population were killed) depressed the birth rate for several years. Reduced kill during the following years, and a change in policy which permitted only males to be killed, allowed the population to recover slowly and return to stability.

Crabeater seals are similar in size and proportions

BAS

Weddell seal mother and pup emerging through midsummer fast ice after a swimming expedition. Females keep holes in the ice open by grinding the edges with their teeth. Canine and incisor teeth are often badly chipped and broken.

Right: Living and feeding among pack ice, crabeater seals are often attacked by killer whales. Many carry three parallel scars on their bodies, telling of an encounter with a killer.

Crabeater seals feed mainly on krill, which they catch in surface waters. The long, down-curved vibrassae or whiskers probably tell them when krill are close at hand.

KOOYMAN

BONE

to Weddells, though usually more slender and considerably more active in their life on the ice. They are the true seals of the pack ice, seldom occurring inshore except in late summer when little of the pack is left. By far the most numerous of Antarctic seals, their total population probably lies between two and five million. Silvery fawn when freshly moulted in spring, they bleach almost to pure white under the summer sun. They suffer more than any other species from attacks of killer whales. In any group of a dozen crabeaters on an ice floe or bergy bit, two or three will inevitably bear the triple razor-slashes of killer whale teeth on their flanks.

In the South Orkney Islands and Antarctic Peninsula region small family groups, consisting of a male, female, and newly-born young, have sometimes been found close inshore during mid-to-late October. Torger Øritsland, a Norwegian biologist who has studied this species among the pack ice of the southern Scotia Arc, found many solitary male-and-female pairs from September onward, and the families remained intact after the pups appeared in mid-September. Later in the year, as far south as McMurdo Sound, pairs with larger pups have been reported, suggesting that the bond remains intact throughout the summer. Young animals moving south in late summer seem often to lose their bearings, sometimes being caught by an early freeze-up and unable to find their way back to the sea. Mummified bodies of young crabeaters have been found many miles inland behind McMurdo Sound, often with clear, undisturbed trails in the sand showing a long, shuffling march before death. In April 1955 some 3,000 young crabeaters were caught by an early freezing of the sea ice near Hope Bay (Antarctic Peninsula). Most of them died of starvation and disease in the ensuing winter.

Crabeaters feed entirely on zooplankton, especially krill; their teeth bear distinctive cusps, which are said to be used in filtering the plankton from the sea (page 138). Their numbers, and the fact that they concentrate near the edge of the sea ice in late winter, make it probable that this species will shortly be exploited commercially. Because it occurs on the high seas, it cannot be protected in any way. Because the species forms pairs, it is extremely likely to be killed in pairs, early in the season when the blubber is thick, and before the young are born. If pelagic sealing in Antarctica begins and proves profitable, the days of the crabeater seal are surely numbered.

CURTSINGER

Leopard seals patrol the waters near penguin colonies, catching the birds as they come in from feeding. The penguins are killed by a bite and skinned neatly by shaking in the water.

KOOYMAN

The business-like teeth of a leopard seal. Leopard seals are aloof, curious animals which follow boats and occasionally swim close to see what is going on. Though frightening, they seldom attack and are easily driven off by shouting.

Leopard and Ross seals *Leopard seals* are also common in the pack ice, but spread widely across the whole of the Southern Ocean and into the Subantarctic as well. Beautiful, slender creatures with silver-grey coats and a formidable array of teeth, they measure up to $3\frac{1}{2}$ metres long. Leopard seals are usually to be found wherever penguins gather, from Cape Crozier and other far southern breeding colonies to Heard Island, Îles Kerguelen and beyond. Usually solitary, they often remain for several months at a time where food is plentiful, patrolling the beaches and catching penguins as they return from fishing. Heard Island has a resident winter population of about 750 animals, the densest concentration reported outside the pack ice. A world population of about 250,000 seems probable.

Within the pack they associate with crabeater seals, possibly feeding on euphausiid plankton among the ice floes. Their teeth are cusped, like those of crabeaters, and are perhaps used in a similar way to filter off the sea water from each mouthful. Leopard seals also eat fish, cephalopods, and possibly the young of other species of seal. Pups are born in November, mostly in the shelter of the pack, though nothing is known of the social organization or family life of this species.

Ross seals are even less known than leopards; up to the 1950s only a very few specimens had been seen or brought back to museums for study. However, icebreakers moving within the pack ice have given biologists the chance to see and collect more of this interesting animal. Slightly smaller than leopard or crabeater seals, Ross seals are dark uniform grey along the back and flanks, silver-grey to white underneath, with a small head, prominent eyes, a short mouth with curious needle-like incisors, and unusually long flippers. They seem to feed mainly on squid, fish and krill. Breeding has not been observed, but Øritsland's notes suggest that pups are likely to be born in November. This is certainly the rarest of the southern seals, with a world population of less than 50,000.

BAS

Ross seal, Antarctica's rarest seal. Confined to pack ice these animals are seldom seen inshore and little is known of their habits or way of life.

Weddell seal hunting among diatom-encrusted floes beneath the sea ice. The large, forward-looking eyes are well adapted for seeing in dim underwater light.

KOOYMAN

Man and the Antarctic

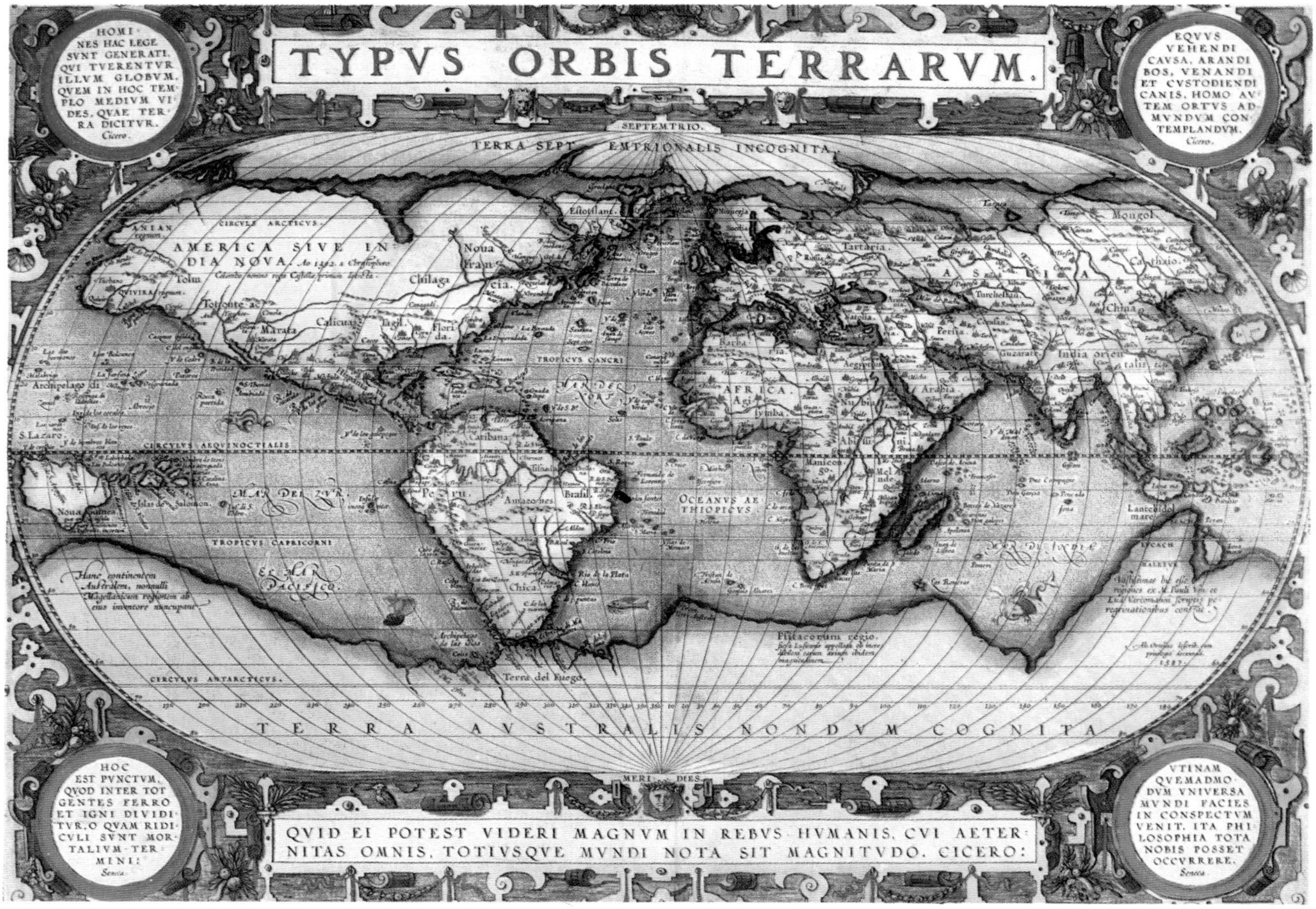

ROYAL GEOGRAPHICAL SOCIETY

Antarctica was discovered by elimination. The huge imaginary southern continent of renaissance map makers, gradually reduced by exploration, eventually became the Antarctic continent of the far south.

Previous page: Man among the penguins. Cape Hallett base, built among a colony of over 60,000 pairs of Adélie penguins in northern Victoria Land.

Man and the northern polar region grew up together. The first human inhabitants of the Arctic pressed northward during a warm interglacial phase some 30,000 years ago, when forest and grassland extended to the polar coast. Later cooling and the spread of the final (Wisconsin-Würm) ice sheets drove them south, but Scandinavian, Siberian and American aboriginals stood ready to move north again when the climate warmed and the ice retreated. A dozen parallel, similar, but independent patterns of human culture arose on the rim of the Arctic basin. Integrating closely with the environment, using its limited resources to the full, but taking no more than he needed, Arctic man evolved like the polar bear, musk-ox and reindeer as part of the Arctic ecosystem.

Man and the Antarctic are comparative strangers, only recently introduced and still ill-at-ease in each others' company. There are no Antarctic aboriginals. Closest to south polar residents were the Indians of Patagonia, whose fires inspired the name Tierra del Fuego for the tip of South America. So far as we know, they did not stir eastward or south, and had no dealings with the Antarctic. There are Polynesian legends of a warrior of the seventh century who sailed far south in his canoe—far enough to see ice on the sea and feel the cold breath of Antarctica. He sailed back as fast as he could, and Polynesian culture spread no further south than the Subantarctic fringe. So the Antarctic region has no human population of its own. Those who discovered the Antarctic were men of the north temperate zone, seeking honour and fortune in an uncomfortable, alien region. Those who followed were exploiters, with every intention of turning the quickest possible profit and getting out.

Discovery

Geographical discovery in the Southern Ocean was a curious reversal of the traditions of exploration. Early voyagers who sailed south across the equator did not discover a new continent. They set out with an enormous continent clearly marked on their charts, and discovered that it was not there. "*Terra australis incognita*" (opposite) was a legacy from early Greek geographers, who knew the earth to be a sphere and believed that a large land mass covered the southern half of it. Renaissance geographers revived the legend, and for over four centuries the great southern continent drew explorers south in search of its wealth. Each voyage drove the elusive coastline back, from tropics to temperate regions, through the zone of westerly winds to the pack ice and the edge of the real Antarctic continent.

Discovery began with Bartholomeu de Novaes, an explorer who in 1488 rounded the southern tip of Africa. Nine years later Vasco da Gama, a fellow Portuguese, sailed eastward round the Cape of Good Hope to Mozambique and India. These voyages between them made Africa a reality, and cleared a vast tract of mythical land from the southern Atlantic and Indian Oceans. In 1520 Fernand Magellan searched southward for a route into the great western ocean which Balboa had seen from Darien. He found it in a winding channel at the tip of South America, with cold, desolate forest on either side and the glow of mysterious fires about him. Contemporary map makers redrew their maps, separating South America from the southern continent, but only by the width of Magellan's strait. Tierra del Fuego became the new continental shore, and remained so until Francis Drake circumnavigated it fifty-eight years later.

Antarctic land first appeared in 1599, when Dirk Gerritsz, Dutch captain of a sailing ship blown southward off his course, saw ice-covered mountains some 500 kilometres south of Cape Horn. Probably the peaks of the South Shetland Islands, their appearance pushed the shores of the unknown continent far below the 60th parallel in this southwestern corner of the Atlantic, and made it clear that any land discovered nearby would barely be worth the finding. In the eastern Indian and Pacific Oceans hope remained longer. Australia was taken to be part of *Terra australis*, from first sighting by Dutch explorers in 1606 to its isolation by Tasman in 1642. New Zealand's west coast, discovered

Îles Kerguelen, discovered in 1772 and believed part of the great southern continent, turned out to be no more than a group of windswept islands. Later they became a base for sealing and whaling. This abandoned whaling station dates from the early twentieth century.

YOUNG

Cook's expeditions to the Southern Ocean took him into fields of pack ice and bergs south of the Antarctic circle.

during Tasman's voyage, in turn became part of the southern continent, and remained so until Cook's circumnavigation over a century later.

In 1728 Jean Bouvet de Lozier, a French officer searching the southern Atlantic for comfortable harbours, discovered a grim, ice-mantled cape close to the Greenwich meridian in 55°S. Inevitably he assumed it to be part of the missing continent, and he sailed for over 3,000 kilometres along the edge of the pack ice in the hope of finding more. Now we know his cape to be part—in fact, practically all—of the forlorn little island which bears his name (page 81). In 1772 the Frenchmen Marion Dufresne and Jacques Kerguelen-Trémarec continued the search for land in the southern Indian Ocean. They discovered Marion Island and Îles Crozet, just north of the Convergence in the African sector. They found also the northern shore of a larger land, green and seemingly fertile, which Kerguelen-Trémarec hailed as part of the missing continent and optimistically called South France. One year later he returned with a company of settlers and domestic animals, to find that his new land, far from being the tip of a rich continent, was yet another dreary island in a cold ocean. Grumbling that he would rather live in Iceland, Kerguelen-Trémarec sailed home in disgust. South France became Îles Kerguelen, now a mid-ocean research station under French administration.

Meanwhile James Cook, of the Royal Navy, had begun the voyages which swept *Terra australis* finally from tropical and temperate latitudes, and disclosed new islands in the Antarctic zone. Among other instructions, Cook was required to search for the southern continent. In his first voyage of 1768–71 he isolated New Zealand and cleared a wide swathe of the temperate southern ocean. His second voyage, of 1772–75, took him further south. In HMS *Resolution*, a converted Whitby collier no bigger than a modern tug, he searched for Bouvet's ice-covered cape, then sailed south to the edge of the pack ice. In January 1773 he crossed the Antarctic Circle for the first time on record, and came almost in sight of Antarctica itself. Returning north to the open sea, he passed close to Îles Kerguelen and headed eastward across the Southern Ocean, keeping close to the 60th parallel. After exploring eastward to the longitude of Tasmania, he turned north and wintered in New Zealand and the warmer Pacific Ocean. Returning south in spring of 1773–74, he crossed the southwestern Pacific in a similar latitude. Late January 1774 found him again south of the Antarctic Circle, this time off Thurston Island, Lesser Antarctica, where he entered the edge of the pack ice and reached the astonishing latitude of 71° 10′S. In almost any other sector of the continent, the same determination and better fortune would have brought him within sight or soundings of Antarctica.

During the third season of his second voyage (January 1775) Cook discovered and circumnavigated South Georgia, landing in Possession Bay. It seemed strange to him that this relatively small island, in a latitude equivalent to that of his home county of Yorkshire, should, ". . . in the very height of summer, be in a manner wholly covered many fathoms deep with frozen snow . . .". He was not impressed by South Georgia: ". . . a savage and horrible country", he wrote, where ". . . not a tree was to be seen, nor a shrub even big enough to make a toothpick". Sailing on before the westerlies, he discovered a line of ice-covered peaks, apparently separated by deep bays, which he called Southern Thule and Sandwich Land. Unable to pass southward, he marked them tentatively as part of a large landmass and sailed on. Today we know them as a chain of isolated volcanic peaks, the South Sandwich Islands (page 70).

Cook was interested in natural history, and carried with him a team of naturalists and artists. Sir Joseph Banks and Dr. D. C. Solander travelled on his first voyage, and two Germans, John Reinhold Forster and his son George, were the observers on his second expedition. Many of the earliest descriptions of polar and subpolar plants date from these remarkable voyages. Cook's own journals, and those of his scientists, also spread word about the wealth of seals in the southern hemisphere, and attracted a grisly industry to Antarctica and the Subantarctic.

Half a century after Cook, a German-born officer of the Imperial Russian Navy completed the elimination of *Terra australis*. Thaddeus von Bellingshausen, a great admirer of the English navigator, planned his two-year voyage to fill in the gaps and uncertainties which Cook's travels had left. In late December 1819 he recharted parts of South Georgia, then sailed eastward to map "Sandwich Land" correctly as a chain of islands. In January 1820 his two ships *Vostok* and *Mirny* crossed the Antarctic Circle and made their way eastward along the fringe of the pack ice. On 28 January, close to the Greenwich meridian, he reported an appearance of land to the south. This was almost certainly the first sighting of the Antarctic continent. Remaining south of Cook's track, he made two further sorties which brought him almost within sight of land again, before turning north and heading across the Indian Ocean for Sydney.

One year later he crossed the southern Pacific Ocean, again managing to keep south of Cook's track for much of the voyage. In January 1821 he discovered Peter I Øy and Alexander Island, off Lesser Antarctica. Von Bellingshausen's voyages finally cut the southern continent down to size, fixing it further south than even the coldest corners of the cold temperate zone.

Heard Island was first visited by sealers in 1855 and rapidly stripped of its fur seals. Elephant seals, subsequently hunted for many years, were boiled down for their oil in iron try-pots, using blubber and skin for fuel.

The sealers

Between the voyages of Cook and von Bellingshausen, European, American and Russian sealers became the explorers of the far south. Their aims were entirely practical. Traditional sealing grounds in the Arctic and Subarctic, for generations over-exploited by ruthless hunting, were showing signs of exhaustion. Southern sealing began about 1778, in the temperate islands off southern South America, Australia and New Zealand, and Cook's newly-discovered Antarctic island of South Georgia. Though other travellers had mentioned seals in their journals and books—often among tall stories of other sea monsters—Cook's accounts and those of his scientists were clearly more than travellers' tales. They reached a wide public, which could well have included enterprising sealing captains and could have been responsible for precipitating the rush to the seal islands.

The first quarry of the sealers was the pelt of the fur seal, for which there were inexhaustible markets in Europe and China. Later the same hunters turned to sea lions and elephant seals, clubbing and boiling them down for their yield of clear oil. Penguins, especially the large king penguin, helped to boost the output of oil on many Subantarctic islands. Hundreds of ships and thousands of men were involved in the sealing industry, which flourished for a frenetic half century before collapsing from exhaustion.

Sealers were the first men to live on South Georgia, Îles Kerguelen, Heard Island, and the South Orkney and South Shetland Islands. Gangs left ashore for the season killed every animal in sight, packing skins and oil into casks to be picked up when their ship returned. They lived dangerously, lowering themselves over cliffs to reach seals on inaccessible beaches, landing through surf on wave-washed reefs to kill wherever and whenever they could. Rival gangs worked together, combining to complete the killing as expeditiously as possible. Turf huts and caves housed them; clothing and supplies were minimal, the rewards small for a season's hard work. Yet they worked with ruthless efficiency, depopulating islands one after another in rapid succession. By 1800 most of the original sealing islands were worked out, and the sealers were beginning to explore southward in their tiny vessels. Rival ships kept their movements secret from each other, and records of some of the earliest voyages of exploration have been lost completely. But the sealers were alert to each others' discoveries, and ready to explore every cove, cranny, beach and headland where seals might be found.

Searching south from New Zealand, the sealers discovered and exploited the Subantarctic Antipodes Islands in 1800, Auckland Islands in 1806, and Campbell and Macquarie Islands in 1810. Each island was cleared of its fur seals within a few seasons. From South Georgia they exploited the South Sandwich Islands, in spite of exposed beaches and hazardous approaches, and relieved Bouvetøya of its small stock of seals. The South Shetland Islands, rediscovered accidentally in February 1819 by a British merchantman blown off course, were visited later in the same year by American, Argentinian and British sealing gangs—who had probably heard news of the new grounds in the waterfront bars of Valparaiso and Montevideo. The following season (1820–21) saw over forty British and American ships working the islands. In 1821–22 the number had increased to over ninety. Within four years of their discovery the South Shetland Islands had yielded over 320,000 fur seal skins and 940 tonnes of oil, and were barely worth a further visit.

STONEHOUSE

Modern sealers worked on South Georgia until 1964, taking elephant seals for oil. The industry, an offshoot of whaling, was closely controlled and did not over-exploit stocks.

Antarctic Peninsula, first seen from the South Shetlands in January 1820, had nothing to offer the sealers who sailed south to investigate. It lay south of the normal range of colonial fur seals and elephant seals, and its native species (leopard, crabeater and Weddell seals) were too sparse and too scattered to be worth hunting. The South Orkney Islands, discovered in 1821, yielded only a few fur seals and a little oil. So the sealers turned explorer, making several remarkable voyages in far southern waters in search of new islands. James Weddell in 1822–23 searched south of the South Orkney Islands, in the sea which now bears his name. He reached 74° 15′S., finding extraordinarily little ice in a region now notorious for its solid pack, and wrote a lively account of southern wildlife, but found no further sealing grounds. The species of inshore seal, which he brought back for the first time to Edinburgh, was named after him. John Biscoe, another British sealer, searched hard for islands to the east of the South Sandwich chain in 1830. Sailing on down the westerlies, he discovered a corner of the Antarctic mainland, which he

MARY EVANS

Ross's royal naval expedition of 1839–42 discovered Victoria Land and the Ross ice shelf (above) and brought back reports that whales were plentiful among the pack ice of the Southern Ocean.

named Enderby Land after his London employers, and crossed the southern Pacific Ocean. In the Peninsula region he discovered Adelaide Island, the Biscoe Islands and parts of the adjoining coast, but he did not discover any new sealing islands. John Balleny, another sealing captain of the firm of Enderby Brothers, discovered the Balleny Islands and, far to the east, Sabrina Land on the continental coast. These voyages made it clear that the sealing bonanza was over.

Oil hunting continued long after the fur sealers had given up. It was a slower business, for carcases were rendered down on the spot in crude iron try-pots. Oiling gangs worked intermittently at Îles Kerguelen, neighbouring Heard Island, Macquarie Island, and probably at many other Antarctic and Subantarctic islands throughout the nineteenth

Dumont D'Urville's discovery of Terre Adélie in 1839. D'Urville landed on Île de la Possession, a small island in a group only a few hundred metres from the continental coast.

MARY EVANS

THE MANSELL COLLECTION

Sperm whales were hunted in tropical and temperate southern waters during the nineteenth century, but were seldom followed into colder, southern seas.

century. Moving from bay to bay, and from island to island, they occasionally found and destroyed small colonies of fur seals left over from the earlier slaughter. During the 1870s there was a revival of interest in the South Shetland and South Sandwich Islands, where fur seals were found to be recovering their numbers. By the end of the decade they had once again been extirpated. Oiling gangs from New Zealand continued to work on Subantarctic Macquarie Island well into the present century, finally ceasing in 1919 after a public outcry. A small elephant sealing industry flourished for many years in South Georgia, as an off-season adjunct to the whaling industry. Organized on rational lines, it maintained a sustained yield over several decades (page 144), apparently without adverse effects on the breeding stocks.

Whaling

Whaling, like sealing, began in the northern hemisphere. The earliest whalers fished off northern European shores in the late sixteenth century, shifted to Svalbard, Greenland and the Bering Sea, and spread into the southern hemisphere during the eighteenth century. Their first prey were the

The first Antarctic factory—whaling station at Gritvyken, South Georgia—opened in 1904. Its catchers scoured the rich seas around South Georgia for nearly sixty years. Now the station lies derelict.

MICKLEBURGH

MERRETT

Harpooning a sea whale from the bow of a whale catcher. The nylon foregoer lead rope snakes out behind the harpoon head, to be followed by a heavy line which will make the whale fast.

Flensing a fin whale on the deck of a factory ship. The lower jaw has been winched away, exposing the baleen plates suspended from the roof of the mouth.

SWITHINBANK

several populations of northern hemisphere right whales and humpback whales. Sperm whales became an important part of the catch during the eighteenth century, and were hunted far into temperate regions of the southern oceans. Antarctic stocks seem to have remained inviolate during these early and middle periods of whaling. Even after the invention of the explosive harpoon it was many years before whalers were tempted into Antarctic waters.

One of the first to comment on the abundance of whales in the far south was James Clark Ross, commander of the British naval expedition of 1839–42, who during his exploration of the Ross Sea found "hunch-back" and sperm whales plentiful as far south as 72°S., and others of a "common black kind—greatly resembling, but said to be distinct from, the Greenland whale". He saw in them "a fresh source of national and individual wealth" awaiting the attention of British commercial enterprise. Enderby Brothers, whose enterprise had sent Biscoe and Balleny south to search for new sealing grounds (page 156) responded promptly by establishing a whaling station on the Subantarctic Auckland Islands. It was not a success; the whales proved difficult to catch and the station was abandoned after two disastrous seasons. In 1892 British commercial enterprise tried again, this time with four old-fashioned whaling ships which sailed from Dundee to the Weddell Sea. Equipped only for the slow-moving humpback, sperm and right whales, which Ross had promised the whalers, they were irritated to find themselves surrounded entirely by fast-moving fin whales and other rorquals, which spouted derisively at them but did not wait to be caught. The *Antarctic*, a Norwegian vessel, similarly equipped and on a similar mission in the Ross Sea, ploughed sadly through schools of rorquals but caught no more than a single minke whale. In compensation, some of its crew made the first landing on the Antarctic continent at Cape Adare (Victoria Land) on 24 January 1895.

Antarctic whaling finally began in 1904, when Argentinian money and Norwegian enterprise combined to establish the first whaling station at Grytviken, South Georgia. Within a decade a dozen factories or factory ships—Norwegian, Chilean, Argentinian and British—were working in South Georgia harbours under licence from the British government, and a further dozen were operating in the sheltered bays of South Shetland and South Orkney Islands. Using explosive harpoons and fast steam catchers, the whalers at first hunted humpbacks close to the islands. When inshore stocks dwindled they turned increasingly to rorquals and hunted further afield. They took only blubber and whalebone, leaving the carcases to float away and rot. In 1912 a floating factory, blocked by ice from its harbour in the South Orkneys, worked for the first time along the edge of the pack ice. After 1925 more and more factory ships took to operating in the open sea, unfettered by licensing and free to hunt wherever the whales were thickest. They spread east of the South Sandwich Islands, and across the southern Indian Ocean. Îles Kerguelen became a whaling centre, and Norwegian hunters penetrated the pack ice of the Ross Sea. Factory ships became bigger and more sophisticated, with slipways for hauling the whales inboard, mechanical strippers and mincers, and plant capable of rendering down the whole carcase. Catchers, too, grew bigger and more efficient. The industry received a boost during the late 1920s, when chemical hardening gave whale oil a new outlet in the manufacture of margarine. In the season 1930–31 whalers achieved a record kill of 40,201 whales, of which almost three in every four were blue whales. The hunting force involved 6 shore stations, 41 factory ships and 232 catchers, mostly British and Norwegian.

The yield of over 600,000 tonnes of oil from this massive slaughter was more than the market could stand in a year of industrial depression, and the price of oil fell. In the following season the Norwegian fleets stayed at home. Thereafter British and Norwegian companies agreed to protect the market by restricting the amount of oil taken in a season, and to restrict the total catch of whalebone whales to a number of "Blue Whale Units" (BWUs). In this calculation one blue whale was considered equal in yield to 2 fin, 2½ humpback or 6 sei whales, and the fleets were free to catch their quota in any convenient combination. They also agreed to limit the season by fixing a formal opening date, so that neither would be tempted to start early while the whales were still thin after their southward migration (page 132). This agreement worked for two seasons. Then the world recovered from its depression and began to re-arm, working up a healthy appetite for oils. Japan, Germany, Panama, South Africa and the United States began whaling during the late 1930s, and the British and Norwegian companies again faced fierce competition. In 1937 a new International Agreement for the Regulation of Whaling was signed by nine nations. This protected right whales (which were no longer a commercial proposition anyway), laid down minimum lengths below which blue, fin, humpback and sperm whales were protected, and restricted the Antarctic pelagic season to three months. The tenth nation, Japan, did not sign.

The 1937–38 season produced a new record catch of 46,039 whales and a yield of over half a million tonnes of oil. The effort involved 2 shore stations, 31 floating factories and 256 catchers. Comparing these figures with those of the 1930–31 record season, it was clear that more whales were now being killed by a smaller and more efficient industrial force, but for a lesser yield of oil. Blue whales, largest of the rorquals (page 130) and always the favourite target of every harpoon gunner, had slipped from three-quarters to less than one-third of the total catch, and many more fin and humpback whales were being taken. The largest Antarctic whales were already disappearing from the ocean.

There was a sharp decline in whaling during the war years. The short respite from hunting did little to help the beleaguered stocks of whales, but the break gave whalers a chance to take note of their own industry. At a wartime meeting in 1944, British, Norwegian and other whalers agreed that future catches of rorquals and humpbacks must be limited if Antarctic stocks were to survive, and an annual catch of 16,000 BWUs was determined. In the post-war season of 1945–46 three British and six Norwegian whaling fleets were in action, and the following year saw the Netherlands, Japan, South Africa and the Soviet Union joining in. A further international meeting in 1946 established the International Convention for the Regulation of Whaling, which led to the development of the International Whaling Commission, and gave Antarctic whaling its own polyglot bureaucracy.

From 1949 the Commission faced the impossible task of controlling a powerful, profitable, highly capitalized, fiercely competitive, multi-national industry—one which had no intention of accepting controls other than on its own terms. Though the terms of reference of the Commission recognized that, under proper management, whales could be harvested rather than quarried, the Commission itself had no powers to impose this view. Though individual nations varied in their approach, collectively whalers saw no alternative to the rapid, efficient exploitation which had proved so successful for the industry in the past. Though the Commission had powers to vary whaling procedure, amendments which tended to restrict whaling were seldom

PRÉVOST

BURNIP

Scott's hut at Cape Evans, McMurdo Sound. Scott's first and second expeditions explored southern Victoria Land and pioneered research in many fields of natural science.

A modern French base on the coast of Terre Adélie: offloading the year's supplies from the expedition ship.

agreed to by all the delegates, and the demurral of one was usually sufficient for the proposal to be dropped by all. So conservation measures which might have maintained the industry in perpetuity, or have postponed its decline, fell by the wayside.

The quota system itself became a trap. Though it set an upper limit to the annual catch, it did not specify how the catch should be made up, or how many factory ships and catchers might be involved in taking it. Thus there was no respite for the large rorquals. The greatest profit fell to those who grabbed the biggest, the most, and the quickest. "The biggest" inevitably meant the large rorquals—the blue and fin whales, which declined rapidly in post-war years (page 128). "The most and the quickest" forced a rat-race between the whaling fleets—a highly competitive standard of efficiency involving crippling capital expenditure. Between 1946 and 1962 the mean tonnage of factory ships increased by almost fifty per cent; catchers almost doubled in tonnage and more than doubled in horsepower. With such heavy costs, the industry could not afford to reduce its annual catch, though it was becoming increasingly apparent that the annual quota of BWUs was much higher than the stocks of humpbacks and rorquals could stand.

Not until 1962 did the Commission secure agreement to divide the quota equably between the nations, so reducing the free-for-all scramble and the pressure on the great whales. But by this time the race was nearly over. South Africa had abandoned pelagic whaling, and the three remaining stations on South Georgia had reached their final season. Britain's fleet was sold to Japan in 1963, the Netherlands' in 1964, and Norway withdrew her last factory ship in 1968. Japan and the Soviet Union, the only whalers left in business, continue to catch on a much reduced quota. The work of the Commission continues. Blue whales, now almost extinct, have joined the list of protected species. Fin whales in their turn have become the gunners' favourite, and, in their turn, have declined in numbers. Sei and sperm whales increasingly form the bulk of the catch, and even tiny minke whales are now considered worth hunting to keep the remnants of the industry afloat. The Blue Whale Unit, appropriately enough, has also disappeared, and each species is hunted to a quota which at last bears some relation to its estimated numbers. But the Commission is still without power, and no species of whale in Antarctic waters is safe from the threat of extermination.

The future of the animals

To the well-adapted polar animal, there is nothing especially difficult or dismaying about the polar environment. Animals do not live in environments to which they are unsuited. The Antarctic environment has selected only a few species from the many which, over the years, have tried to invade it, and has moulded them to its stringent but tolerable requirements. As a relatively new environment, it is still selecting, and some of the species described in this book may still have some way to go before they are as fully adapted as others. But by any ecological standards the animals of the Antarctic region are no less successful, in their day-to-day lives, than many animals of temperate and tropical environments.

As specialized animals, rigorously matched to a rigorous environment, polar species are vulnerable to environmental changes. As isolated animals in a relatively empty homeland, they may be more than usually sensitive to disturbance by other species. As animals of a simple, unbuffered ecosystem they are peculiarly liable to suffer if any of their delicately balanced relationships with other species are upset. On each of these counts they are vulnerable to man—a large noisy animal with larger, noisier machines, who changes environments to suit himself, disturbs everything with his restless excess of energy, and plays havoc with ecosystems wherever he goes. Even when he is not butchering seals to make fur coats, or whales to make lipstick, margarine and pet foods, his very presence in Antarctica, with ships, aircraft, tractors, dogs, is a disturbance to the local inhabitants.

To his credit, man is sometimes aware of his disruptive influence and tries to protect the animals he is disturbing. In Antarctica he has tried hard. The land areas of the southern polar regions are better provided than many civilized countries with legislation to protect native flora and fauna against unnecessary disturbance. Except for a single quadrant which nobody claims, every square metre of continental Antarctica and every island south of the Convergence is claimed by an interested patron—Norway, Australia, France, New Zealand, Britain, Chile or the Argentine. Each claimant nation has framed regulations, usually sensible, to protect the plants and animals in its territory. So for instance, every Adélie penguin breeding on the coast of Antarctica is protected by the laws of territory in which she nests. On Antarctic Peninsula she is protected by three sets of laws, for Chile,

MICKLEBURGH

British Antarctic Survey Base "H", Signy Island, a small modern research station. Fireproof living quarters and laboratories house the scientists and technicians who man the station continuously, mostly on one-year or two-year contracts.

SWITHINBANK

Annual supplies for a British base include seal carcases, mostly crab-eaters, to feed the dog teams.

Counting elephant seals in a snowstorm. Seal research involves surveys and counts of populations, often in difficult weather conditions.

BURLEY

Britain and the Argentine each claim and guard her nesting ground.

More important, she has the nominal goodwill of a dozen nations. The seven claimants, together with five other nations (U.S.S.R., U.S.A., Belgium, Japan and South Africa), have signed the Antarctic Treaty, which dedicates Antarctica and all lands within the 60th parallel to peaceful international uses. The Treaty, ratified in 1961 and in force for thirty years, lists among its objectives the preservation and conservation of living resources, and all the signatory nations subscribe to a code, *Agreed measures for the conservation of Antarctic flora and fauna*, which outlines practical ways in which expedition members and tourists may reduce their impact on the polar ecosystem. A biological working party of S.C.A.R. (Scientific Committee on Antarctic Research), on which the nations are represented by polar biologists, recommends, from time to time, special conservation measures to be adopted, helps to promote research, and keeps governments aware of their responsibilities under the Treaty. Not all the nations are equally concerned about conservation, but there is strong general concern to make the Treaty work, and the animals of the Antarctic benefit from it. The Treaty also bans weapons testing, nuclear explosions and the disposal of radioactive wastes in Antarctica. The Antarctic Treaty, by showing concern for the strange biological communities within the 60th south parallel, has led to an awareness among scientists, administrators and casual visitors that the continent and islands hold something precious which must not be destroyed for trivial reasons.

Unfortunately, the Treaty does not cover the high seas or the pack ice within the Antarctic region. These areas were specifically excluded from its provisions, as the legislators felt that their inclusion would raise problems which might prove insuperable and endanger the Treaty as a whole. So the Antarctic Ocean, which holds practically all of the biological resources of the region, remains unprotected by any international agreement. Though most of its animals are at present of no economic interest, the pelagic seals—those which breed and spend most of their lives on the pack ice—are especially vulnerable to commercial exploitation and could well be man's next Antarctic prey.

Legally, the pelagic seals have no more protection than the fur seals and elephant seals of the nineteenth century and the rorquals of the twentieth century. However, a small committee of seal biologists, working through S.C.A.R. and the

CURTSINGER

Gentoo penguin and radio transmitter involved in US research project on homing and orientation behaviour.

Research on inshore fishes helps biologists to unravel ecological relationships in polar waters.

EDWARDS

MANNERING

Helicopters operating from ships have helped scientists to land on many otherwise inaccessible islands. US navy helicopter over the Balleny Islands.

MANNERING

US icebreaker alongside the ice foot at McMurdo Station, the main US base in Antarctica. Mast and huts of the base camp are visible on the ridge in the background. Cargo vessels bring supplies to this point, less than 1,500 kilometres from the South Pole.

Rubbish disposal is a constant problem in Antarctica where natural decay is slow and packaging accumulates.

SWITHINBANK

Hercules transport of the US Navy, fitted with skis for flights between main and outlying stations.

SWITHINBANK

MICKLEBURGH

Splendidly protected by their warm fur against all but the very worst weather, dogs are still used as sledge-haulers in Antarctica. They can eat dried meat but fresh seal carcases are always available.

Soviet tractor train, Lazarev ice shelf. Tracked vehicles with sledges are used for long journeys across central Antarctica and allow scientists to take heavy equipment into the field.

SWITHINBANK

Treaty organization, has drawn up recommendations for the voluntary regulation of a pelagic sealing industry, suggesting annual catch limits for each species (with complete immunity for one, the Ross seal), and other measures which would protect the stocks against over-exploitation. If these measures are accepted by the Treaty nations, it is just possible that crabeater, leopard and Ross seals will be saved, for the Treaty nations are those most likely to become involved in sealing operations. However, this can only be a voluntary control. If the costs of sealing operations force hunters to increase their catches beyond the recommended limits, or if the limits are found to be optimistically high, it will be difficult indeed for the Treaty nations to keep control of the situation. And there is nothing to prevent hunters from non-Treaty nations taking all they want at any time.

A more diffuse problem awaits international legislators when they turn their attentions to Antarctic krill. The large crustaceans of the plankton are present in quantities so enormous that, sooner or later, man will begin seriously to harvest them for their protein and fat content. Almost certainly they will become an object of competition; when one nation can process them satisfactorily, every advanced nation with capital to spend will want to join in the krill bonanza. The rorquals are estimated to have taken some 50 million tonnes of krill annually from the sea in their heyday. If remaining stocks of whales now consume one-fifth of that amount, and if other predators have not increased to take the place of the missing whales, then some 40 million tonnes of krill could be available for harvesting each year. As Soviet scientists have pointed out, this compares with the weight of fish which man at present takes each year from the sea. It would not take modern technology long to build up a powerful, profitable, highly capitalized, fiercely competitive multi-national industry—just like the whaling industry of the 1950s and 1960s—to snatch this third and final harvest from the Antarctic Ocean.

One can only hope that, in the next few critical years, the common sense which forged the Antarctic Treaty will be applied to the more revolutionary task of forging an international agreement for Antarctic marine resources. An agreement which allowed man to draw rationally on the bounty of Antarctic waters would ensure both a lasting and generous supply of proteins and oils for man, and a lasting place in the polar environment for the animals of Antarctica.

Bibliography

AUSTIN O L Jr. (ed.): *Antarctic Bird Studies*
Antarctic Research Series 12, American Geophysical Union, Washington DC, 1968

BAKAYEV, V G: *Atlas Antarktikii*
Main Administration of Geodesy and Cartography, Ministry of Geology, Moscow, 1966

H M STATIONERY OFFICE: *Final act of the Conferences on Antarctica together with the Antarctic Treaty, etc.*
Cmnd. 913, Misc. No. 21, London, 1959

HART, T J: *Phytoplankton Periodicity in Antarctic Surface Waters*
Discovery Reports 21: 261–336, Cambridge University Press, 1942

HATHERTON, T (ed.): *Antarctica*
Methuen, London, 1965

HOLDGATE, M W (ed.): *Antarctic Ecology* (2 vols.)
Academic Press, London and New York, 1970

KING, J E: *Seals of the World*
British Museum, London, 1964

MACKINTOSH, N A: *The Stocks of Whales*
Fishing News, London, 1965

ORVIG, S (ed.): *Climates of the Polar Regions*
World Survey of Climatology 14, Elsevier, Amsterdam, 1970

PREVOST, J and MOUGIN, J L: *Guide des Oiseaux et Mammifères des Terres Australes et Antarctiques Françaises*
Delachaux et Niestlé, Paris, 1971

PRIESTLEY, R, ADIE, R J and ROBIN, G de Q (eds.): *Antarctic Research*
Butterworths, London, 1964

QUAM, L O (ed.): *Research in the Antarctic*
American Association for the Advancement of Science, Publication 93, Washington DC, 1971

SMALL, G L: *The Blue Whale*
Columbia University Press, London and New York, 1971

WATSON, G E and others: *Birds of the Antarctic and Subantarctic*
Folio 14, Antarctic Map Folio Series, American Geographical Society of New York, 1971

Glossary

Ahumic (Of soil) containing little or no humus or organic material.

Amphibolite A metamorphic rock containing amphibole (a silicate mineral) and plagioclase felspar; usually unfoliated.

Amphipod A group of widely distributed crustacean animals, including sand hoppers, "shrimps" of the suborder Gammaridae, and many planktonic forms which swarm in polar seas.

Baleen "Whalebone": horn-like plates which grow from the roof of the mouth of Mysticete whales, forming a filter which separates tiny food animals from sea water.

Biota The living animals and plants of a region.

Culminicorn Horny plate covering the ridge of the bill in penguins, petrels and other birds.

Crustacean Member of a class of invertebrate animals with hard shell, jointed limbs and (usually) gills: includes crabs, shrimps, euphausiids, etc.

Diatom Single-celled plant with siliceous shell, common in surface waters of polar regions in summer. Very important as the main photosynthesizers or "fixers" of solar energy.

Ecosystem The **biota** (see above) of an area together with its habitat or place of living.

Ectoparasite A parasite which lives externally on its host, e.g. flea, louse.

Euphausiid **Crustacean** (see above) of the family Euphausiidae, many species of which form huge shoals in Antarctic surface waters, and are eaten by whales, seals and other carnivores.

Flensing Taking the blubber off a whale carcase.

Fumarole Volcanic vent, through which gases escape to the open air.

Geosyncline Trench or valley in the earth's crust, usually submarine, into which sediments settle over long periods.

Globigerina Single-celled animals with chalky shells, which live in surface waters of the ocean.

Gneiss Banded or foliated rock, usually igneous or metamorphic.

Humic (Of soil) containing a high proportion of humus or organic material.

Intrusion Mass of rock injected while liquid into other rock formations, subsequently cooling to form distinctive layers or masses.

Katabatic (Of wind) flowing downhill in response to gravity, e.g. masses of cold, dense air flowing down the seaward slopes of the Antarctic continent, forming strong local winds.

Monotreme Egg-laying mammal (e.g. platypus, spiny ant-eater).

Metamorphic (Of rock) altered by intense heating, folding, etc.

Ornithogenic (Of soil) containing a high proportion of bird droppings.

Obliquity Tilt of the earth's axis, i.e. its angle in relation to the plane in which the earth rotates about the sun.

Pelagic Living at the surface of the sea.

Photosynthesis The process by which plants absorb the energy of the sun, linking carbon dioxide and water (in the presence of the green pigment chlorophyll) to form glucose.

Phytoplankton The plants which float in the surface waters of the ocean (e.g. **diatoms**; see above).

Piedmont Ice sheet formed usually by the fusion of glaciers along the lower slopes and foothills of a mountain range.

Polynya Lake of open water in the middle of an ice sheet.

Schist Split or foliated rock, usually one which has been altered by **metamorphic** processes (see above).

Spermaceti Waxy oil contained in the "casket" or forehead of a sperm whale; formerly much in demand for candle-making.

Tuff Soft fibrous rock exploded from a volcano.

Zooplankton The small animals which drift in the surface waters of the ocean (e.g. **Euphausiids**; see above).

Index

Numbers in italics refer to illustrations.